Introduction to Geophysics

A Student Handbook

Ndikum Eric Ndoh

DEDICATION

I heartily dedicate this book to God Almighty, the principal Author of Geophysics.

Foreword

The Earth holds an abundance of wealth beneath its surface. From the vast reserves of oil and gas to the intricate details of ancient civilizations, accessing this wealth requires specialized tools and techniques. Geophysics as an applied science amongst others has over the years contributed immensely to the unveiling of this wealth beneath the Earth's surface.

This book, "Introduction to Applied Geophysics, A Student Handbook" serves as a valuable guide for those embarking on their journey into this fascinating field. If you are a student seeking a comprehensive foundation in the subject or simply someone with a curiosity about the surface of the Earth beneath our feet, this book offers a clear and engaging introduction to the principles and applications of applied geophysics.

Through its well-structured chapters, the book explores the fundamental concepts of gravity, magnetism, electricity, and seismics, laying the groundwork for understanding how these concepts can be used to map the Earth's interior. The knowledge is further exemplified with the use of solved examples on practical elements which deepens the understanding the concepts presented. It then moves into the practical aspects of data acquisition and interpretation, equipping readers with the tools to translate the whispers of the Earth into meaningful geological information. Finally, the book showcases the diverse applications of geophysics in fields ranging from resource exploration and infrastructure development to environmental monitoring and archaeological investigations.

"Introduction to Applied Geophysics, A Student Handbook" is an excellent illustration to the power of geophysics in revealing the unseen forces that shape our planet. As you prepare to move through its pages, prepare to be captivated by the potential of this science to picture the earth's interior.

Prof. TABOD Charles TABOD

Professor of Geophysics and Dean of the Faculty of Science

The University of Bamenda

Preface

Welcome to the exciting world of applied geophysics! This book is designed to be your companion on a journey of exploration, where we will delve into the methods and techniques used to probe the Earth's concealed wonders.

This book is structured in a way that progressively builds your understanding. Part 1 provides a solid foundation in the fundamental principles of physics that govern geophysical measurements. Part 2 equips you with the tools to acquire and interpret geophysical data, transforming raw numbers into meaningful insights about the subsurface. Part 3 showcases the diverse applications of geophysics across various disciplines, demonstrating the real-world impact of this science. Finally, Part 4 proceeds into more advanced topics, offering a glimpse into the cutting-edge techniques used by professionals in the field (Optional).

Throughout the book, you will encounter real-world case studies that illustrate the power of geophysics in solving practical problems. Each chapter offers clear explanations, helpful diagrams, and suggestions for further exploration. The content draws inspiration from suggested materials you might already possess, making this book a valuable resource for students and those who are simply curious about the discipline of Geophysics alike.

As you embark on this journey, keep your curiosity alive and ask questions! Geophysics is a dynamic field with constant advancements. This book may not provide all the answers, but hopefully, it will spark your passion for exploring the earth's sub-surface and inspire you to forge deeper into the fascinating world beneath our feet.

ACKNOWLEDGMENTS

I sincerely appreciate Prof. Tabod Charles for accepting to do the forward of this book and for all the insightful contributions which improved the quality of the book

I also acknowledge KDP for the free images used on the cover pages.

Part 1: Foundations of Geophysics

Chapter 1: Introduction to Geophysics

1.1 The Realm of Geophysics

Our planet Earth, seemingly solid and familiar on the surface, harbors a vast and complex interior shrouded in limited knowledge. From the presence of buried mineral resources to intricate geological formations and vital water resources, the resources beneath our feet hold immense significance and influence our lives in numerous ways. **Geophysics**, at the intersection of physics and geology, emerges as a powerful tool for identifying these resources and having a better picture of the subsurface layers of our planet (Telford et al., 1999).

This fascinating field employs a diverse arsenal of techniques, each drawing upon fundamental principles from physics, to gather crucial data about the Earth's interior.

These principles include:

- **Gravity:** By analyzing variations in the Earth's gravitational field, geophysicists can gain insights into the distribution of mass within the planet, revealing information about the structure and composition of the crust, mantle, and core (Reynolds, 2011). This analysis often involves applying principles of Newton's law of universal gravitation, which states that the gravitational force (F) between two objects is directly proportional to the product of their masses (M_1 and M_2) and inversely proportional to the square of the distance (r) separating them:

$$F = \frac{GM_1M_2}{r^2} \tag{1}$$

where G is the gravitational constant and has value 6.6743×10^{-11} N·m²/kg².

- **Magnetism:** The Earth possesses a natural magnetic field generated by the motion of molten iron in its outer core. This magnetic field interacts with certain minerals, particularly those containing iron, nickel, or other magnetic elements (Lowrie, 2007). By studying variations in the Earth's magnetic field and how it interacts with different materials, geophysicists can gain valuable insights into the presence and location of mineral deposits, geological structures, and even ancient human settlements.

- **Electricity:** The ability of a material to conduct electricity, known as **electrical conductivity**, varies significantly depending on its composition and water content. By injecting electrical currents into the ground and analyzing the resulting electrical response, geophysicists can map variations in subsurface conductivity, aiding in the identification of groundwater aquifers, potential mineral deposits, and even environmental contamination plumes (Chen et al., 2024). This is based mostly on the relationship between electrical current (I), voltage (V), and resistance (R) through Ohm's Law:

$$V = I\ R \tag{2}$$

- **Seismic Waves:** The Earth's interior is not static, but constantly vibrates with various types of seismic waves generated by earthquakes, volcanic eruptions, and even human activities. These waves travel through different rock layers at varying speeds depending on the density and elasticity of the rock. By analyzing the **arrival times, amplitudes, and travel paths** of these waves, geophysicists can create detailed images of the Earth's interior, similar to how medical ultrasound uses sound waves to image the human body (Sheriff and Geldart, 1995).

Wielding these tools, geophysicists act as detectives, piecing together the complex puzzle of the Earth's realms below the surface. They contribute to various crucial endeavors, including:

1. Imaging and Mapping the Earth's Interior: By utilizing various geophysical techniques, scientists can create detailed **geological cross-sections** (Figure 1.1: Earth's Internal Structure) that reveal the structure and composition of the Earth's interior, including the crust, mantle, and core. This knowledge is fundamental for understanding the Earth's formation, evolution, and the processes shaping our planet (Lowrie, 2007).

2. Locating and Characterizing Valuable Resources: Determining the presence and location of valuable resources like metals, oil, and gas is crucial for sustainable resource exploration and development. Geophysical methods play a central role in this process by helping identify potential areas for exploration and characterizing the properties of these resources (Parasnis, 1986).

3. Identifying and Assessing Geological Hazards: Earthquakes, landslides, and volcanic eruptions pose significant threats to life and property. By analyzing various geophysical data, scientists can identify areas susceptible to these hazards and assess their potential impact, aiding in disaster mitigation and preparedness efforts (Dobrin and Savit, 1988).

4. Investigating and Monitoring Environmental Challenges: The ever-growing concern for environmental protection necessitates the development of effective monitoring and remediation strategies. Geophysical methods offer valuable tools for investigating and monitoring environmental challenges like groundwater contamination and soil remediation, enabling informed decision-making towards a healthier planet (Chen et al., 2024).

5. Understanding the History of the Earth: The Earth's history is intricately woven into the layers of rock beneath our feet. By studying the physical properties and characteristics of these rocks through geophysical methods, scientists can gain valuable insights into past geological events and processes, providing a deeper understanding of our planet's journey through time (Sheriff and Geldart, 1995).

In conclusion, geophysics serves as a powerful bridge between the visible surface of our planet and the sub-surfaces within. By harnessing the principles of physics and employing sophisticated techniques, geophysicists act as explorers, venturing into the unseen depths of the Earth. Their discoveries and insights contribute significantly to various fields, from resource exploration and environmental monitoring to hazard assessment and understanding the Earth's history. As technology continues to advance, the capabilities of geophysical methods are constantly evolving, promising even deeper and more comprehensive insights into the fascinating world beneath our feet.

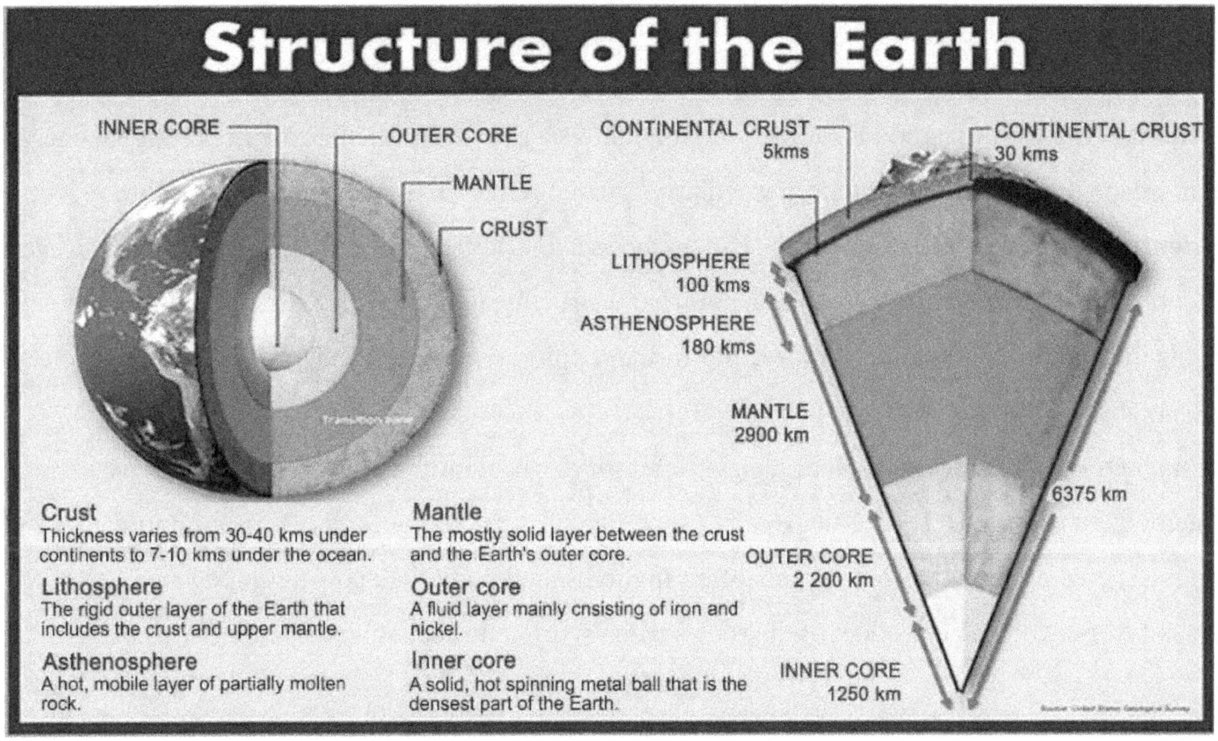

Figure 1.1: Earth's Internal Structure (Compass, https://compass.rauias.com/geography/internal-structure-earth-crust-mantle-core-discontinuities/ of 7 March, 2024)

1.2 Applications of Geophysics

The realm of geophysics extends far beyond the initial exploration and discovery of the Earth's treasures. The information gleaned from various geophysical techniques plays a crucial role in various sectors, influencing our present and shaping our future.

Let's now proceed into some noteworthy applications:

1. Resource Exploration: Discovering Nature's Bounty

- **Minerals:** The insatiable human desire for infrastructure development, energy production, and technological advancements necessitates the discovery and responsible extraction of essential minerals. Geophysics plays a vital role in this quest by locating and characterizing deposits through various techniques like the magnetic surveys, electrical resistivity measurements and seismic reflection profiling with which geophysicists can identify potential areas containing valuable minerals like copper (Cu), gold (Au), and iron (Fe). This information guides further exploration and resource extraction efforts (Reynolds, 2011).

- **Oil and Gas:** Hydrocarbon resources, like oil and gas, play a significant role in our global energy landscape. Geophysical methods, particularly seismic reflection surveys, are crucial for:
  - **Identifying potential reservoirs:** By analyzing the reflection patterns of seismic waves, geophysicists can identify potential underground structures that may contain oil and gas deposits. This information is essential for directing further exploration efforts and ensuring energy security (Nabighian, 1991).
  - **Characterizing reservoirs:** Once a potential reservoir is identified, geophysicists can use seismic methods to estimate the size, shape, and properties of the reservoir, allowing for a more efficient and targeted extraction process. This often involves analyzing the acoustic impedance (Z) of the rock layers, which is a product of the rock's density (ρ) and seismic wave velocity (V):

$$Z = \rho V \quad (3)$$

- **Groundwater:** Access to clean and sustainable water resources is vital for human life and societal well-being. Geophysics plays a crucial role in mapping and characterizing aquifers in which electrical resistivity surveys and seismic methods are often used to map the extent and properties of underground water-bearing layers, known as aquifers. This information is vital for sustainable water resource management and ensuring access to clean water for drinking, irrigation, and industrial uses (Rubin & Hubbard, 2005).

2. Environmental and Engineering Applications: Safeguarding Our Planet and Infrastructure

- **Geohazards:** The Earth's dynamic nature can sometimes present challenges in the form of natural disasters like earthquakes, landslides, and volcanic eruptions. Geophysics plays a crucial role in:
  - **Assessing geohazard risk:** By analyzing various geophysical data, such as seismic activity, ground deformation, and magnetic anomalies, geophysicists can assess the potential risk and severity of these hazards. This information is vital for implementing mitigation strategies like building earthquake-resistant structures, establishing early warning systems, and planning evacuation procedures (Keating and Fraser, 2009).
- **Environmental Monitoring:** Protecting our environment necessitates continuous monitoring and assessment. Geophysics offers valuable tools for:

- **Investigating and monitoring environmental concerns:** Electrical resistivity surveys, seismic refraction surveys, and other methods can be used to investigate and monitor issues like groundwater contamination, soil pollution, and land subsidence. This information helps guide remediation efforts and enables informed decision-making towards environmental protection (Burger, 1992).

- **Engineering Projects:** Constructing safe and sustainable infrastructure requires a thorough understanding of the subsurface conditions. Geophysics aids in:
  - **Locating buried utilities:** By utilizing electromagnetic methods, geophysicists can locate and map buried utilities like pipelines and cables, preventing accidental damage during construction projects (Telford et al., 1999).
  - **Assessing foundation stability:** Seismic surveys and electrical resistivity measurements can be used to assess the stability of slopes and foundations for various construction projects, ensuring the safety and longevity of structures.
  - **Investigating potential hazards:** Geophysical surveys can be used to identify potential hazards like sinkholes, cavities, or faults in the subsurface, allowing engineers to design and implement appropriate mitigation strategies.

3. Scientific Research: Studying the Earth's interior and Shaping Our Future

Geophysics plays a vital role in advancing our scientific understanding of the Earth, contributing to:

- **Understanding Earth's formation and evolution:** By studying the Earth's internal structure and composition through various geophysical techniques, scientists can gain insights into the planet's formation, evolution, and the processes that have shaped it over time (Menke, 2012).
- **Studying past climate changes:** Analyzing the physical properties of rocks through geophysical methods can provide valuable information about past climate changes, enabling scientists to develop models for understanding future climate trends and mitigate the impact of climate change.
- **Investigating tectonic processes:** Studying seismic activity and other geophysical phenomena helps us understand the movement and interaction of tectonic plates, contributing to a deeper understanding of the forces shaping the Earth's surface and influencing the distribution of resources and natural hazards.

Conclusion

By getting a better understanding of the interior of the Earth, geophysics plays a vital role not only in resource exploration and environmental protection but also in shaping our future through sustainable development and scientific advancements. This chapter has provided a glimpse into the diverse and impactful world of geophysics, laying the foundation for your exploration of its exciting applications in the chapters to come. As technology and our understanding of the Earth continue to evolve, the possibilities and applications of geophysics are seemingly limitless, promising even greater contributions to building a safer, more sustainable, and knowledge-driven future for generations to come.

Examples and exercises

1.1 The Realm of Geophysics

Example 1.1: Imagine you are an archaeologist searching for a buried city. You wouldn't just start digging at random. Instead, you might use tools like ground-penetrating radar or aerial photography to identify potential areas of interest. Similarly, geophysicists use various techniques to "see" beneath the Earth's surface without directly digging.

Example 1.2: Imagine you're a doctor trying to diagnose a patient without directly observing their organs. You might use X-rays, CT scans, or other imaging techniques to gain insights into their internal structure. Similarly, geophysicists utilize various techniques to "see" beneath the Earth's surface, drawing inferences about its composition and structure without direct physical excavation.

Exercise 1.1:

- List three physical properties of Earth that geophysicists can measure to learn about the subsurface.

Solution 1.1:

1. Gravity: Variations in Earth's gravitational field can reveal the presence of denser or less dense materials.
2. Magnetism: Magnetic anomalies can indicate the presence of certain minerals or geological features.
3. Electrical conductivity: Differences in electrical conductivity can help map groundwater, locate mineral deposits, and identify environmental contamination.

Exercise 1.2:

Imagine you're a geophysicist investigating a potentially oil-rich region. Identify three different geophysical methods you could use to gather data about the subsurface and explain how each method assists in your investigation.

Solution 1.2:

1. **Seismic Reflection Survey:** By generating controlled seismic waves and analyzing their reflections from different Earth layers, we can create detailed images of the subsurface, revealing the thickness and composition of rock formations. This information can help identify potential oil reservoirs based on the characteristic reflection patterns associated with oil-bearing rocks.
2. **Gravity Survey:** Measuring subtle variations in the Earth's gravitational field can provide clues about the distribution of mass within the subsurface. Since oil generally has a lower density than surrounding rock, identifying areas with lower gravity might indicate the presence of potential oil reservoirs.
3. **Magnetic Survey:** Analyzing the Earth's magnetic field and its interaction with the subsurface can help locate magnetic minerals often associated with specific rock formations, sometimes even providing clues about the presence of oil deposits depending on the geological context.

1.2 Applications of Geophysics

Example 1.3: Imagine a team of engineers needs to build a new bridge. They would need to know the type of soil and rock beneath the surface to ensure the bridge is stable. Geophysicists can use seismic surveys to map the underground layers and help choose the safest and most cost-effective location for the bridge.

Example 1.4: Imagine you're an environmental scientist concerned about potential groundwater contamination from a nearby industrial facility. You could collaborate with a geophysicist to utilize electrical resistivity surveys. These surveys measure the ability of the subsurface to conduct electricity, which can vary depending on factors like water content and the presence of contaminants. This information can help map the extent and potential severity of the contamination, aiding in remediation efforts.

Exercise 1.3:

- Match the following applications of geophysics with their corresponding descriptions:
  - A. Resource Exploration
  - B. Environmental Monitoring
  - C. Geohazards
  - D. Scientific Research

1. Studying past climate changes and tectonic processes.
2. Locating and characterizing valuable mineral deposits.
3. Assessing the risk of earthquakes, landslides, and volcanic eruptions.
4. Investigating and monitoring groundwater contamination.

Solution 1.3:

1. D
2. A
3. C
4. B

Exercise 1.4:

Geophysics plays a crucial role in various sectors. Match the following applications with their corresponding descriptions and explain the specific geophysical techniques typically employed in each case:

Applications:

- A. Resource Exploration
- B. Engineering Projects
- C. Geohazards
- D. Environmental Monitoring

Descriptions:

1. Assessing the risk of earthquakes by analyzing seismic activity and ground deformation patterns. (___, Explain Technique)
2. Mapping the extent and thickness of groundwater aquifers using electrical resistivity surveys and seismic methods. (___, Explain Technique)
3. Locating buried utilities like pipelines and cables to prevent accidental damage during construction projects. (___, Explain Technique)
4. Identifying potential mineral deposits through magnetic surveys and electrical resistivity measurements. (___, Explain Technique)

Solution 1.4:

1. **C. Geohazards**

   Explanation:

- **Seismic Reflection and Refraction Surveys:** These techniques map the subsurface structures and measure seismic wave velocities, which can be used to assess fault locations, potential earthquake epicenters, and ground stability.

2. **D. Environmental Monitoring**

   Explanation:

- **Electrical Resistivity Surveys:** By measuring the electrical conductivity variations in the subsurface, these surveys can help identify potential contaminant plumes due to differences in conductivity compared to surrounding groundwater.

3. **B. Engineering Projects**

   Explanation:

- **Ground Penetrating Radar (GPR) and Electromagnetic Surveys:** These methods utilize electromagnetic waves to detect and map the location of buried objects like pipes and cables with varying conductivity and permittivity compared to the surrounding soil.

4. **A. Resource Exploration**

   Explanation:

- **Magnetic Surveys:** Measuring variations in the Earth's magnetic field can help locate magnetic minerals often associated with specific valuable mineral deposits.
- **Electrical Resistivity Surveys:** These surveys can also be used in conjunction with magnetic surveys to further characterize the properties of potential mineral deposits.

Chapter 2: Physical Principles for Geophysical Exploration

Obtaining knowledge of the contents of the Earth's subsurface requires the application of various physical principles. This chapter presents the fundamental concepts behind three key geophysical methods: gravity, magnetism, and electricity, exploring how they aid in our quest to understand and map the subsurface.

2.1 Gravity and the Earth's Interior

Gravity, the invisible force binding us to the Earth, holds immense power in revealing the components within our planet. This section explores how variations in Earth's gravitational field, caused by its uneven mass distribution, become a powerful tool for geophysical exploration.

The Uneven Pull: Understanding Gravity Anomalies

The Earth's interior is not a uniform sphere of equal density. Instead, it is layered, with varying densities from the solid crust to the molten outer core and the solid inner core (Blakely, 1995). This **density variation** creates an **unequal gravitational pull** across Earth's surface, leading to **gravity anomalies** (Reynolds, 2011).

These anomalies represent **subtle deviations** from the average gravitational acceleration of around 9.8 m/s^2. They can be **positive** (higher than average) or **negative** (lower than average) depending on the underlying material.

Unveiling the Earth's Interior: The Power of Gravity Measurements

Geophysicists utilize sophisticated instruments called **gravimeters** to measure these minute gravity variations with incredible precision. These instruments operate on the principle behind Newton's law of universal gravitation, which defines the **gravitational force (F)** (see equation 1).

By analyzing gravity anomalies, we can estimate the **mass distribution** within the Earth. This, in turn, helps us **image** the distribution of denser and less dense materials at different depths, creating a kind of **gravitational picture** of the Earth's interior (Blakely, 1995).

For instance, **positive anomalies** can often indicate the presence of **denser materials** like metallic ores or thicker crustal sections (**Figure 2.1**). Conversely, **negative anomalies** might hint at **less dense features** like sedimentary basins or salt domes (Blakely, 1995).

By studying gravity anomalies, we can unlock valuable insights into what lies beneath the Earth's surface, making gravity surveys a vital tool for several applications:

1. Imaging the Earth's Internal Structure:

Beyond qualitative interpretations, advanced techniques can leverage the quantitative information obtained from gravity measurements. These techniques involve complex mathematical models and calculations based on physics principles, allowing for a more quantitative understanding of the Earth's internal structure (Blakely, 1995).

For example, the **inverse problem** approach aims to find a 3D mass distribution within the Earth that best explains the observed gravity anomalies. This process often involves iterative methods and sophisticated software to find the most plausible solution.

2. Locating Potential Mineral Deposits: Gravity's Guiding Hand

Dense metallic ores and sulfide minerals typically have a higher density than surrounding rocks. This difference in density often results in **positive gravity anomalies**, providing valuable clues for **mineral exploration** (Parasnis, 2006). By analyzing these anomalies, geophysicists can target areas with a higher probability of containing valuable mineral deposits, significantly increasing exploration efficiency.

3. Identifying Hydrocarbon Reservoirs: A Subtle, yet Insightful Approach

While the density variations between oil and gas and surrounding rock are subtler, they can sometimes be detected through gravity surveys, especially when **combined with other geophysical methods** like seismic surveys (Telford et al., 1990). This **combined approach** can aid in **identifying potential hydrocarbon reservoirs** for further exploration and resource extraction.

In conclusion, gravity measurements act as a powerful tool for studying the depths of the Earth. By analyzing variations in the gravitational field, we gain valuable insights into the planet's internal structure, resource potential, and even geological processes. This knowledge plays a crucial role in various fields, from resource exploration and geological mapping to understanding the Earth's formation and evolution. As technology advances, the sensitivity and precision of gravity measurements are constantly improving, promising even greater insights into the Earth's interior and the potentials within it.

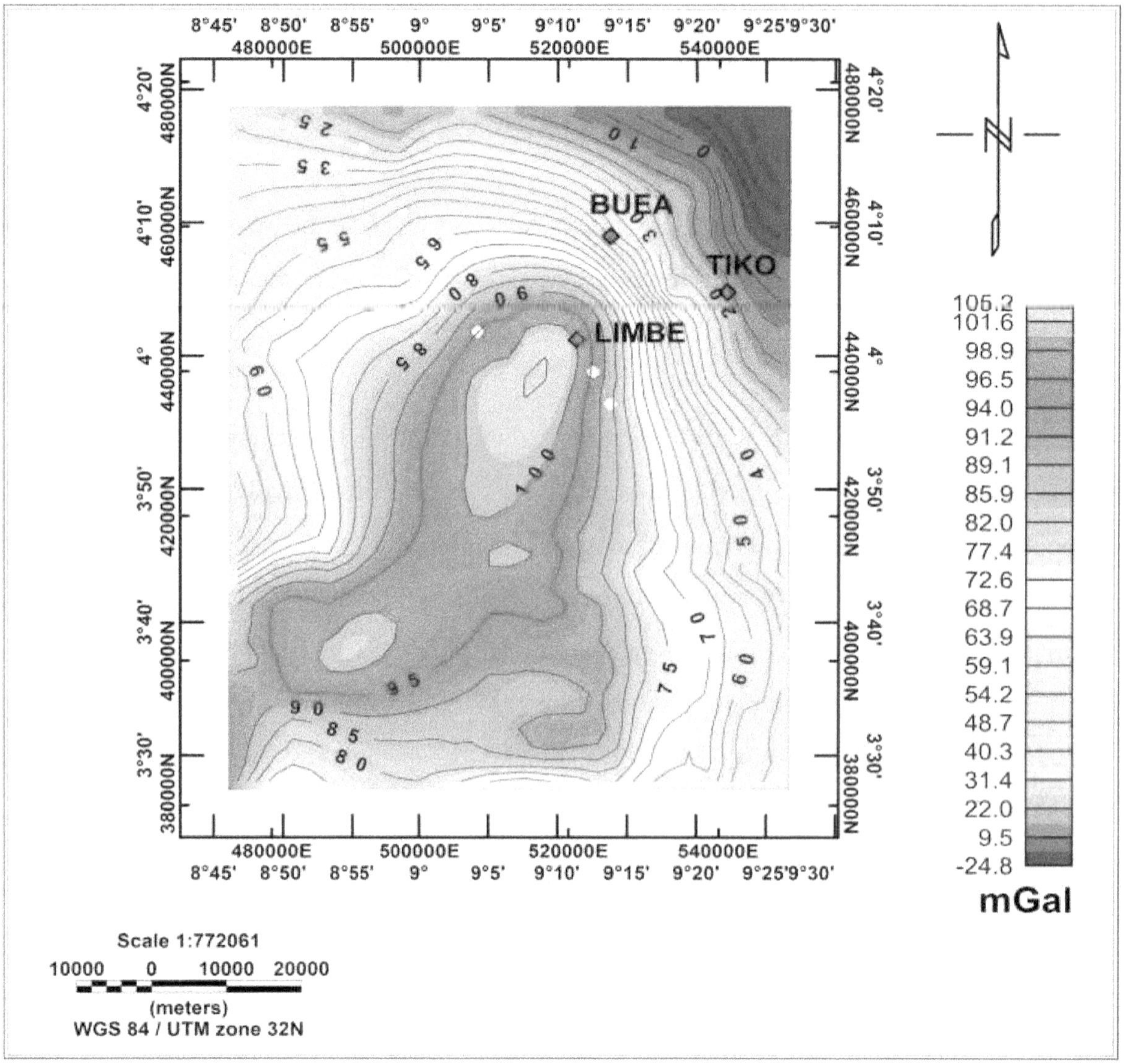

Figure 2.1: Gravity anomaly map of the Douala Sedimentary basin which illustrates the distribution of gravity anomalies and highlights regions with positive and negative deviations. (Ndikum et al., 2019)

2.2 Magnetism and Mineral Exploration

Our planet is cloaked in a natural magnetic field, generated by the churning motion of molten iron in its outer core. This magnetic field isn't just a passive shield against harmful solar radiation; it also interacts with certain minerals, particularly those containing iron, nickel, or other magnetic elements (Lowrie, 2007). This interaction forms the foundation for **magnetic surveys**, a powerful tool for mineral exploration and geological investigations.

The Power of Magnetic Interaction: Exploring the Earth's Surface

Magnetic surveys involve measuring the Earth's magnetic field and its variations at different locations on the surface. This measurement is often done using instruments called **magnetometers**, which can be ground-based, airborne, or even marine.

The Earth's magnetic field is not uniform, and variations in this field can be caused by several factors, including:

- **Magnetic minerals:** Minerals like magnetite (Fe_3O_4) and hematite (Fe_2O_3) are strongly magnetic, meaning they can significantly alter the local magnetic field in their vicinity.
- **Geological structures:** Faults, dikes, and other geological features can have contrasting magnetic properties compared to surrounding rocks, leading to observable variations in the magnetic field (Reynolds, 2011).
- **Man-made objects:** Buried objects like pipelines, metallic debris, or even archaeological remains can also cause local anomalies in the Earth's magnetic field.

By analyzing these variations, we can unlock valuable insights into what lies beneath the Earth's surface, making magnetic surveys a vital tool for several applications:

1. Mineral Exploration: A Magnetic Compass for Finding Treasures

Magnetic minerals like magnetite and hematite create distinct **magnetic anomalies** in the Earth's magnetic field. These anomalies can be positive (higher than the surrounding field) or negative (lower than the surrounding field), depending on the type and amount of magnetic material present (Keating and Fraser, 2009).

By interpreting these anomalies, geophysicists can identify potential locations for valuable mineral deposits. For example, a strong positive anomaly might indicate the presence of iron ore deposits, while a negative anomaly could suggest a pocket of non-magnetic minerals like diamonds or kimberlite pipes.

2. Mapping Geological Structures: Unveiling the Sub-surface Architecture

Magnetic anomalies can sometimes reveal the presence of **geological structures** with contrasting magnetic properties compared to surrounding rocks. These structures can include:

- **Faults:** These are fractures in the Earth's crust where rocks have shifted. Depending on the type of rock on either side of the fault, a fault can create either a positive or negative magnetic anomaly (Reynolds, 2011).
- **Dikes:** These are igneous intrusions that have solidified within existing rock formations. Their magnetic signature can differ from the surrounding rock, making them detectable through magnetic surveys.
- **Basement rocks:** These are the underlying rocks beneath sedimentary layers. By analyzing magnetic anomalies, geologists can map the extent and depth of these basement rocks, providing valuable insights into the region's geological history.

3. Archaeological Investigations: Unearthing the Past

Surprisingly, buried structures from the past can also leave their mark on the Earth's magnetic field. Objects like:

- **Kilns and hearths:** Used for heating and cooking, these structures often contain fired clay and iron oxides, which can create weak magnetic anomalies.
- **Weapons and tools:** Made from iron or other magnetic materials, these artifacts can cause subtle magnetic anomalies, especially when clustered together.

By analyzing these weak magnetic signatures, archaeologists can identify potential locations for buried settlements, ancient structures, and even lost artifacts, aiding in their excavation and historical research (Clark and Clark, 1990).

In conclusion, magnetic surveys, through their ability to detect variations in the Earth's magnetic field, offer a powerful and non-invasive tool for exploring the potentials and features beneath the surface. From identifying valuable mineral deposits and mapping geological structures to unearthing the understanding of the past, magnetic surveys play a crucial role in various fields, contributing significantly to our understanding of the Earth's history and composition.

2.3 Electricity and Subsurface Conductivity

The ability of a material to conduct electricity, known as **electrical conductivity**, varies significantly depending on its composition and water content. Materials like metals, due to their free-flowing electrons, are excellent conductors, while insulators like dry rock offer significant resistance to current flow. This fundamental principle forms the basis for **electrical and electromagnetic (EM) methods** of geophysical exploration.

The Language of Electricity: Measuring the Earth's Response

These methods involve injecting controlled electrical currents into the ground through electrodes and measuring the resulting **electrical response** at different locations on the surface. This response can be measured in various ways, depending on the specific technique employed. Some methods directly measure the **electrical resistivity**, which is a measure of a material's resistance to current flow, following the formula:

$$\rho = \frac{VA}{IL} \tag{4}$$

where:

- ρ is the resistivity
- **V** is the measured voltage difference between two electrodes
- **I** is the injected current
- **A** is the area of the current-carrying electrode
- **L** is the distance between the current-injecting and voltage-measuring electrodes

Other techniques may measure the induced **electromagnetic field** generated by the current flow within the subsurface.

By analyzing these electrical or electromagnetic responses, geophysicists can gain valuable insights into the properties and characteristics of the subsurface, unlocking information about:

1. Subsurface Conductivity Variations: A Map of Earth's Conductivity

Different geological formations exhibit varying electrical conductivities. For instance, **clay-rich soils** and **saline groundwater** are typically good conductors, while **dry rock** and **air** are poor conductors. This variation in conductivity allows us to:

- **Map groundwater aquifers:** Groundwater-filled zones generally have higher conductivity compared to surrounding rock, making them detectable through electrical resistivity surveys (Jin, 1992).
- **Locate potential mineral deposits:** Some metallic ore deposits, such as those containing sulfides or copper, can exhibit higher conductivity compared to surrounding rock, aiding in their exploration (Ali, 2023).
- **Identify environmental contamination plumes:** Pollutants like heavy metals or saline solutions can increase the conductivity of the subsurface, allowing for the potential detection of environmental contamination (Jin, 1992).

2. Soil Properties: Building a Solid Foundation

Electrical resistivity surveys are often used in **engineering applications** to investigate soil properties for various purposes, including:

- **Determining foundation stability:** By mapping variations in subsurface resistivity, engineers can assess the suitability of a site for construction projects by identifying areas with weak or unstable soil (Ali, 2023).
- **Locating buried utilities:** Electrical cables and pipelines can sometimes be detected through their contrasting conductivity compared to surrounding soil, aiding in their location and avoiding accidental damage during excavation projects.

3. Hydrocarbon Exploration: A Complementary Approach

While less common than other applications, specific EM methods, like **magnetotelluric (MT) surveys**, can be used to map the resistivity variations associated with hydrocarbon reservoirs. These variations, although subtle, can sometimes provide complementary information when used in conjunction with other geophysical data like seismic surveys.

In conclusion, electrical and electromagnetic methods, through their ability to measure and interpret variations in the Earth's electrical conductivity, offer a powerful tool for non-invasively investigating the subsurface. These methods contribute significantly to various fields, from resource exploration and environmental monitoring to engineering construction and understanding the Earth's geological makeup. As technology advances, the sensitivity and resolution of electrical and EM methods continue to improve, promising even deeper insights into the sub-surface of the earth.

2.4 Seismic Waves and Imaging Earth's Layers

The Earth's interior, despite its apparent stillness, is a dynamic environment constantly vibrating with various types of seismic waves. These waves, generated by earthquakes, volcanic eruptions, and even human activities like explosions, travel through the Earth at different speeds. This crucial difference in speed is governed by the **density (ρ)** and **elasticity** of the rock layers through which they propagate. By analyzing the **arrival times, amplitudes, and travel paths** of these waves, geophysicists can gain a wealth of information about the Earth's interior, obtaining understanding about it layer by layer.

The Language of Seismic Waves: Decoding the Earth's Symphony

Seismic waves come in two main categories: **body waves** and **surface waves**.

- **Body waves:** These waves travel through the Earth's interior, and they are further classified into two types:
  - **P-waves (primary waves):** These are the **fastest** seismic waves, and they can travel through both solids and liquids. They vibrate in a **compressional** manner, meaning they alternately squeeze and stretch the material they travel through. Their speed (V_p) is related to the material's **bulk modulus (K)**, shear modulus (μ) and density (ρ) according to the formula:

$$V_p = \sqrt{\left(\frac{K + 4/3\,\mu}{\rho}\right)} \qquad (5)$$

In fluids (liquids and gases), the shear modulus is zero, so the p-wave velocity becomes:

$$V_p = \sqrt{\left(\frac{K}{\rho}\right)} \qquad (6)$$

  - **S-waves (secondary waves):** These waves are **slower** than P-waves and can only travel through **solids**. They vibrate in a **transverse** manner, meaning they move the particles perpendicular to the direction of wave propagation. Their speed (V_s) is related to the material's **shear modulus (μ)** and density (ρ) according to the formula:

$$V_S = \sqrt{\frac{\mu}{\rho}} \tag{7}$$

- **Surface waves:** These waves travel along the Earth's surface and are typically slower than body waves. They come in two main types:
    - **Love waves:** These surface waves vibrate horizontally, similar to S-waves.
    - **Rayleigh waves:** These surface waves are the **slowest** seismic waves and involve a combination of vertical and horizontal movement of the ground particles.

By understanding the properties of these different seismic wave types and their interaction with various rock layers, geophysicists can utilize them for various purposes:

1. Imaging the Earth's Internal Structure: A Seismic Portrait

One of the most powerful applications of seismic waves lies in **seismic reflection**, a technique analogous to medical ultrasound. By generating controlled seismic waves using sources like explosions or vibrators and recording their reflections from different Earth layers, geophysicists can create detailed images of the Earth's interior (Sheriff and Geldart, 1995).

These images, visualized in **seismic reflection profiles** (Figure 2.2), reveal the **thickness, composition, and boundaries** of various layers, including the crust, mantle, and core. This information is crucial for understanding the Earth's structure, evolution, and the processes shaping our planet.

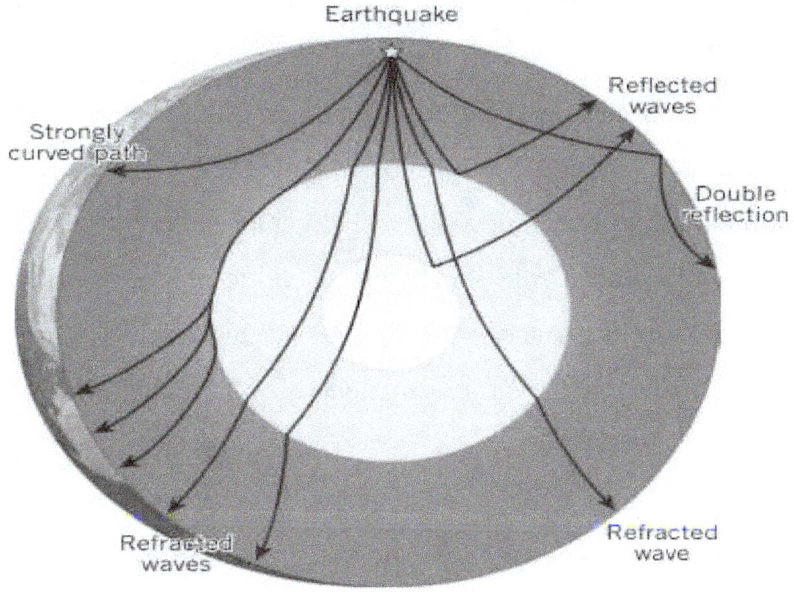

Figure 2.2: Seismic Reflection Image of Earth's Layers (Quora of 07/03/2024; https://qph.cf2.quoracdn.net/main-qimg-b8dd07b2174119009de55c713f674068-lq)

2. Mapping Geological Structures: Unveiling the Architecture of the Earth

Seismic waves can be invaluable in **mapping geological structures** like faults, folds, and intrusive rocks. These structures can significantly impact resource exploration and geohazard assessment.

- **Faults:** When seismic waves encounter a fault, they can be reflected, refracted, or diffracted, depending on the fault's orientation and displacement. Analyzing these wave behavior changes allows geophysicists to identify the location and characteristics of faults (Dobrin and Savit, 1988).
- **Folds:** Folds in rock layers can also alter the path and speed of seismic waves. By studying these variations, geophysicists can map the location and geometry of folds, providing insights into the region's geological history and potential resource distribution.

3. Locating Oil and Gas Reservoirs: A Seismic Guide

Seismic reflection is the primary method used for **exploration and characterization of oil and gas reservoirs**. Subtle differences in the seismic properties of rocks containing oil and gas compared to surrounding rock, even though small, can be detected through careful analysis of seismic reflection data.

By identifying these subtle variations, geophysicists can help pinpoint potential locations for drilling and resource extraction, playing a critical role in meeting the world's energy demands.
In conclusion, seismic waves, through their unique properties and interaction with the Earth's interior, offer a powerful tool for getting more knowledge of the Earth's interior. From revealing the deeper layers of the Earth to mapping geological structures and locating valuable resources, seismic exploration plays a crucial role in various scientific and engineering endeavors. As technology advances, the resolution and sensitivity of seismic methods continue to improve, promising even deeper insights into the intricate symphony of the Earth's interior.

Understanding the principles of these fundamental geophysical methods paves the way for further exploration in subsequent chapters, delving into how these techniques are applied in various real-world scenarios.

Examples and exercises

2.1 Gravity and the Earth's Interior

Example 2.1: Imagine you are weighing yourself on a scale at different locations on Earth. You might notice slight variations in your weight due to subtle differences in Earth's gravitational pull. These variations, although small, can be detected by sensitive instruments like gravimeters and used by geophysicists to map variations in density within the Earth.

Example 2.2: Imagine you're a detective investigating a unknown compartment in a room. While you can't see directly inside, you discover subtle weight variations in the floorboards when walking over different areas. Similarly, geophysicists use gravimeters, incredibly sensitive instruments, to measure tiny variations in Earth's gravity. These variations, though slight, offer clues about the density of materials beneath the surface.

Exercise 2.1:

- Explain how a positive gravity anomaly might be interpreted by a geophysicist exploring for mineral deposits.

Solution 2.1:

A positive gravity anomaly indicates a region with denser than average material compared to the surrounding rock. This could be interpreted by a geophysicist as a potential location for mineral deposits, as many valuable minerals, like iron ore, are denser than the surrounding rock. However, other factors like variations in rock types and local geology should be considered to confirm the presence of minerals.

Exercise 2.2:

Imagine you're a geophysicist analyzing a region with a **negative gravity anomaly**. How might you interpret this finding, and what additional information would you seek to solidify your interpretation?

Solution 2.2:

A negative gravity anomaly indicates a region with **less dense material** compared to the surrounding rock. This could be interpreted in several ways:

- **Presence of voids or cavities:** Less dense areas could be due to air-filled spaces, caverns, or even ancient lava tubes.
- **Igneous intrusions:** Certain types of igneous rocks, formed from cooling and solidifying magma, can be less dense than the surrounding rock, potentially causing a negative anomaly.
- **Sedimentary basins:** Sedimentary rocks, often less dense than their parent rock, can accumulate in basins, leading to a negative anomaly.

However, relying solely on a gravity anomaly is insufficient for definitive conclusions. To solidify your interpretation, you might seek additional information like:

- **Geological maps:** Existing geological maps can provide insights into the region's rock types and potential geological formations that could explain the anomaly.
- **Seismic surveys:** Seismic waves travel faster through denser materials. Analyzing seismic data can help confirm the presence of low-density layers or voids identified through the gravity anomaly.
- **Magnetic surveys:** Magnetic anomalies can sometimes provide complementary information about the types of rock present in the region, aiding in refining the interpretation.

By combining data from various geophysical techniques with geological knowledge, geophysicists can gain a more complete understanding of the subsurface and its characteristics.

2.2 Magnetism and Mineral Exploration

Example 2.3: Imagine you have a magnet and it attracts certain objects, like iron nails, but not others, like plastic toys. Similarly, Earth's natural magnetic field interacts with specific minerals containing iron, nickel, or other magnetic elements. Geophysicists use magnetic surveys to measure these interactions and identify potential locations for these valuable minerals.

Example 2.4: Imagine you're searching for buried treasure chests on a beach using a metal detector. While the detector helps locate metallic objects, it might also pick up signals from other metal objects like bottle caps or even nails. Similarly, magnetic surveys for mineral exploration, while valuable, have limitations that need to be considered.

Exercise 2.3:

- Describe two limitations of using magnetic surveys for mineral exploration.

Solution 2.3:

1. **Magnetic minerals are not the only sources of magnetic anomalies:** Other geological features like certain rock formations can also cause magnetic anomalies, making it challenging to definitively identify the presence of valuable minerals solely based on magnetic surveys.
2. **Magnetic signals can be attenuated by certain materials:** Dense materials like overlying rock layers can weaken the magnetic signal, making it difficult to detect anomalies from deeper mineral deposits.

Exercise 2.4:

Describe two alternative geophysical methods, besides magnetic surveys, that could be employed in conjunction with magnetics for a more comprehensive mineral exploration strategy. Briefly explain the principle behind each method and how it can complement magnetic surveys.

Solution 2.4:

1. Electrical Resistivity Surveys:

- **Principle:** This method measures the ability of the subsurface to conduct electricity. Different minerals and rock types have varying electrical conductivity.
- **Complementing Magnetics:** Electrical resistivity surveys can help distinguish between different types of conductive minerals and differentiate them from non-conductive magnetic anomalies, providing a more specific indication of potential mineral deposits.

2. Gravity Surveys:

- **Principle:** This method measures variations in Earth's gravity, which are influenced by the density of the subsurface materials.
- **Complementing Magnetics:** While not directly detecting specific minerals, gravity surveys can help identify density anomalies that might be associated with certain types of mineral deposits, further refining the exploration targets based on the combined information.

By employing various geophysical methods in conjunction with magnetic surveys, we gain a more comprehensive understanding of the subsurface properties and can potentially pinpoint mineral deposits with greater accuracy and confidence.

These exercises showcase the importance of considering limitations and incorporating complementary techniques for robust and successful geophysical exploration.

2.3 Electricity and Subsurface Conductivity

Example 2.5: Imagine you have a garden hose. Water flows easily through a wide-open hose but encounters resistance when the hose is kinked or narrowed. Similarly, the ability of the Earth's subsurface to conduct electricity varies depending on its composition and water content. By measuring this electrical resistance, geophysicists can gain insights into the properties of the subsurface.

Example 2.6: Imagine you're a plumber trying to locate a leak in a buried pipe network. You might use specialized tools that send electrical signals through the pipes and analyze the signal variations to identify the location of the leak. Similarly, geophysicists use electrical resistivity surveys, which involve injecting electrical currents into the ground and analyzing the resulting response, to gain insights into the properties of the subsurface.

Exercise 2.5:

- Explain how electrical resistivity surveys can be used to help locate groundwater.

Solution 2.5:

Groundwater is generally a good conductor of electricity compared to dry rock or soil. By measuring electrical resistivity variations through electrical resistivity surveys, geophysicists can identify zones with lower resistivity, potentially indicating the presence of groundwater. This information can be crucial for locating and managing water resources.

Exercise 2.6:

Explain how electrical resistivity surveys can be used not only for locating groundwater but also for environmental monitoring purposes.

Solution 2.6:

While electrical resistivity surveys are valuable for locating groundwater due to its higher conductivity compared to dry soil or rock, they also hold immense potential for environmental monitoring:

- **Identifying Contaminant Plumes:** Certain contaminants, like pollutants from industrial waste or saltwater intrusion in coastal regions, can have different electrical conductivity compared to the surrounding groundwater. By analyzing variations in resistivity, geophysicists can map the extent and potential severity of these contaminant plumes, aiding in remediation efforts.
- **Monitoring Landfill Sites:** Electrical resistivity surveys can be used to monitor landfill sites for potential leachate migration. Leachate, the liquid that percolates through landfill waste, often has a higher conductivity than the surrounding soil, and its movement can be tracked through resistivity variations, enabling early detection of potential environmental concerns.

By analyzing the electrical resistivity of the subsurface, geophysicists can gain valuable insights not only for locating valuable resources like groundwater but also for safeguarding our environment through effective monitoring and remediation strategies.

2.4 Seismic Waves and Imaging Earth's Layers

Example 2.7: Imagine throwing a pebble into a pond. You observe ripples or waves spreading outwards from the point of impact. Similarly, earthquakes and other events generate seismic waves that travel through the Earth's interior at different speeds depending on the density and elasticity of the rock layers they encounter. By analyzing these waves, geophysicists can "image" the Earth's internal structure.

Example 2.8: Imagine you're an archaeologist trying to understand the layout of a buried city. You might use ground-penetrating radar, which sends and analyzes radio waves, to create an image of the structures beneath the surface. Similarly, geophysicists use seismic waves, generated by earthquakes or controlled sources, to "see" through the Earth's interior and understand its composition and structure.

Exercise 2.7:

- Describe the difference between seismic reflection and seismic refraction methods.

Solution 2.7:

- **Seismic reflection:** This method involves analyzing the reflection of seismic waves from different layers within the Earth. Reflected waves indicate boundaries between layers with contrasting properties. This technique is analogous to using sound waves to image internal organs in medical ultrasound.
- **Seismic refraction:** This method involves analyzing how seismic waves bend or refract as they travel through layers with varying velocities. By studying the refraction patterns, geophysicists can determine the thickness and properties of different layers within the Earth.

These examples and exercises provide a basic understanding of the physical principles behind some key geophysical methods. Remember that these are simplified explanations, and real-world applications involve complex techniques and interpretations.

Exercise 2.8:

Imagine you're a geophysicist studying a specific region using both seismic reflection and seismic refraction methods. Explain how the information obtained from each method can be combined to create a more comprehensive picture of the subsurface.

Solution 2.8:

Seismic Reflection:

- Provides high-resolution images of the subsurface, revealing details about the **boundaries and layering** of different rock formations.
- Think of it like a detailed map with precise locations of different features.

Seismic Refraction:

- Determines the **thickness and elastic properties** of different layers within the Earth.
- Think of it like understanding the depth and composition of each layer marked on the map.

By combining the information from both methods:

- We can correlate the **reflections** observed in seismic reflection data with specific **boundaries** identified through seismic refraction, allowing for a more accurate interpretation of the rock types present.
- We can use the **thickness and velocity data** from seismic refraction to **calibrate** the depth scale of the seismic reflection image, providing a more accurate representation of the subsurface structure.

Therefore, combining these methods provides a **richer and more comprehensive understanding** of the Earth's interior, enabling geophysicists to accurately map the subsurface and gain valuable insights into its composition and structure.

Chapter 3: Mathematical Tools for Geophysical Analysis

Just as a skilled detective utilizes various tools to unravel a mystery, geophysicists rely on a powerful arsenal of mathematical and computational techniques to decipher the understanding of layers beneath the Earth's surface. This chapter presents the essential mathematical tools that form the foundation of geophysical analysis.

3.1 Basic Mathematics and Physics Fundamentals

A successful exploration of the Earth's sub-surface depths necessitates a strong foundation in mathematics and physics. These fundamental disciplines provide the tools necessary to understand the physical principles underlying various geophysical phenomena and interpret the data gleaned from diverse instruments and techniques.

Mathematics:

- **Algebra:** The ability to manipulate and solve equations involving various physical parameters is paramount. From calculating gravitational attraction using Newton's Law of Universal Gravitation to analyzing the relationship between electrical current (I), voltage (V), and resistance (R) through Ohm's Law, algebra forms the backbone of countless geophysical calculations (Reynolds, 2011).
- **Trigonometry:** Analyzing the spatial relationships between points and objects plays a crucial role in interpreting geophysical data. Whether it's calculating the dip and strike of geological formations based on field observations or determining the location of a buried object using the principles of triangulation, a solid understanding of trigonometric concepts is essential (Blakely, 1995).
- **Calculus:** The powerful tools of differentiation and integration find numerous applications in geophysics. From analyzing the propagation of seismic waves using wave equations (involving second-order partial derivatives) to modeling the flow of groundwater through porous media using Darcy's Law (which incorporates calculus concepts like partial derivatives and integrals), mastering calculus allows geophysicists to tackle complex problems and model various physical processes (Lowrie, 2007).

Physics:

Beyond the realm of mathematics, a thorough grasp of fundamental physics principles is crucial for interpreting diverse geophysical phenomena:

- **Mechanics:** Understanding the principles of motion, forces, and energy is essential for analyzing the behavior of seismic waves, studying the movement of tectonic plates, and even interpreting gravity anomalies associated with variations in mass distribution within the Earth.
- **Electricity and Magnetism:** The concepts of electric and magnetic fields, their interactions with matter, and the principles of electromagnetic induction play a vital role in various geophysical methods. From interpreting magnetic anomalies associated with specific minerals to understanding the principles behind electrical resistivity surveys, these concepts are indispensable for geophysicists (Telford et al., 1990).

In conclusion, mastering the language of mathematics and physics equips geophysicists with the necessary tools to not only understand the fundamental principles governing the Earth's internal processes but also to analyze and interpret the vast amount of data collected through various methods. This strong foundation forms the cornerstone for successful exploration, resource discovery, and ultimately, a deeper understanding of our planet's dynamics and history.

3.2 Calculus and Differential Equations in Geophysics

Calculus, particularly the realm of differential equations, emerges as a powerful tool in various geophysical applications, allowing us to model and analyze complex Earth processes.

1. Wave Propagation and Seismic Exploration:

At the heart of seismic exploration lies the ability to model the propagation of **seismic waves** through the Earth's interior. These waves, generated by earthquakes or controlled sources, carry valuable information about the subsurface structure and properties. Understanding their behavior and interpreting the data they carry necessitates the application of **wave equations**.

One such example is the **acoustic wave equation** for an isotropic (having the same properties in all directions) elastic medium, given by:

$$\frac{\partial^2 t}{\partial^2 t} = (V_P^2 + V_S^2)\,\nabla^2 u \qquad (8)$$

where:

- $u(x, y, z, t)$ represents the wave displacement at a specific location (x, y, z) and time (t)
- V_p and V_s are the compressional (P-wave) and shear (S-wave) wave velocities, respectively
- ∇^2 is the Laplacian operator, representing the sum of second-order partial derivatives with respect to spatial coordinates (x, y, z)

Solving this differential equation, considering the specific initial and boundary conditions of a particular seismic event, allows geophysicists to model the propagation of seismic waves through various Earth layers with varying properties. This modeling capability forms the foundation for interpreting seismic data, understanding the Earth's internal structure, and locating potential resources (Sheriff and Geldart, 1995).

2. Gravity Modeling and Mass Distribution:

Another crucial application of differential equations lies in **gravity modeling**. Earth's gravitational field exhibits slight variations due to the uneven distribution of mass within the planet. Analyzing these variations is essential for understanding the Earth's internal structure and identifying potential resource deposits.

To model and interpret these gravity variations, geophysicists often utilize **Poisson's equation**:

$$\nabla^2\Phi = -4\pi G\rho \tag{9}$$

where:

- Φ is the gravitational potential
- G is the gravitational constant
- ρ is the mass density at a specific point

By solving this differential equation for the gravitational potential Φ, considering the known or estimated density distribution (ρ) within the Earth, geophysicists can gain insights into the subsurface mass distribution and its potential influence on the observed gravity anomalies (Blakely, 1995).

3. Electromagnetic Field Modeling and Subsurface Exploration:

The principles of electromagnetism also find various applications in geophysics. For instance, **electromagnetic induction** plays a crucial role in numerous surveying techniques used to map the subsurface conductivity and identify potential resources. Understanding the behavior of electromagnetic waves used in these surveys often involves solving **Maxwell's equations**, a set of four fundamental equations governing the interaction between electric and magnetic fields:

- $\nabla \cdot E = \frac{\rho}{\varepsilon_0}$ (Gauss's law for electricity) (10)
- $\nabla \cdot B = 0$ (Gauss's law for magnetism) (11)
- $\nabla \times E = -\frac{\partial B}{\partial t}$ (Faraday's law of induction) (12)
- $\nabla \times B = \mu_0(J + \varepsilon_0\frac{\partial E}{\partial t})$ (Ampère's law with Maxwell's addition) (13)

where:
E is the electric field
B is the magnetic field
ρ is the electric charge density
ε_0 is the permittivity of free space
μ_0 is the permeability of free space
J is the electric current density

While solving Maxwell's equations in their entirety can be complex, simplifying these equations based on specific assumptions and applying them to particular geophysical scenarios allows geophysicists to model the behavior of electromagnetic waves and interpret the data collected from various surveys.

In conclusion, differential equations act as powerful tools in the geophysicist's arsenal, enabling the modeling and analysis of diverse Earth processes like wave propagation, gravity variations, and electromagnetic interaction with the subsurface. This capability paves the way for a deeper understanding of our planet's structure, resource exploration, and ultimately, a comprehensive picture of Earth's dynamic geological processes.

3.3 Computer Programming for Data Processing

In the modern era of geophysics, the ability to leverage computational power through computer programming has become an indispensable skill. Mastering programming languages like Python, MATLAB, or C++ equips geophysicists with the tools to navigate the vast and complex realm of geophysical data, transforming it into meaningful insights and driving advancements across various fields.

1. Automating Complex Calculations and Minimizing Errors:

Geophysical data analysis often involves repetitive calculations and intricate mathematical manipulations. Tasks like inverting seismic data to infer subsurface properties, processing and filtering large datasets from various sensors, and solving complex differential equations governing Earth processes can be time-consuming and prone to human error. By harnessing the power of programming, geophysicists can automate these tasks.

For instance, a Python script can be written to implement a specific filtering algorithm, iteratively applying the filter to each data point in a large seismic dataset. This not only saves valuable time but also reduces the potential for errors introduced through manual calculations (Burger, 1992).

2. Visualizing and Analyzing Large Datasets: Effective Storytelling with Data

Geophysical datasets often reach immense sizes, encompassing vast amounts of information collected from various instruments and techniques. Analyzing and extracting meaningful insights from such data requires sophisticated visualization tools. Programming languages like Python and MATLAB offer extensive libraries and frameworks specifically designed for creating compelling and informative visualizations.

By writing code, geophysicists can create:

- **Interactive plots** that allow users to explore data trends and relationships from different perspectives.
- **3D visualizations** that provide a more comprehensive understanding of complex spatial relationships within the data.
- **Custom maps** that integrate various datasets, offering a holistic view of the investigated region.

These visualizations not only aid in data analysis but also serve as powerful communication tools, enabling geophysicists to effectively share their findings with collaborators, stakeholders, and the broader scientific community.

3. Pushing the Boundaries: Developing Custom Models and Algorithms for Specialized Applications

While numerous software packages cater to general geophysical data processing and analysis, the field constantly evolves, demanding innovative approaches to address specific challenges and unlock new possibilities. Programming empowers advanced users to:

- **Develop custom models:** By writing code, geophysicists can implement novel theoretical models tailored to specific research questions or unique geophysical settings. For instance, one might develop a custom model to simulate the behavior of electromagnetic waves in a complex geological environment.
- **Create specialized algorithms:** Programming allows for the creation of efficient and tailored algorithms for specific data processing tasks or geophysical applications. This can involve developing innovative signal processing techniques, noise reduction algorithms, or even advanced inversion methods to extract even deeper insights from existing data (Menke, 2012).

In conclusion, mastering computer programming equips geophysicists with the essential tools to navigate the ever-growing realm of data. From automating complex tasks and effectively visualizing information to pushing the boundaries through custom models and algorithms, programming empowers geophysicists to get a better understanding of our planet, paving the way for advancements in resource exploration, environmental monitoring, and scientific discovery.

Examples and exercises

3.1 Basic Mathematics and Physics Fundamentals

Example 3.1: Imagine you are investigating the strength of a bridge. You might use the principles of mechanics to calculate the forces acting on the bridge and ensure its stability. Similarly, geophysicists use basic mathematics and physics principles to understand the behavior of various geophysical phenomena, like the propagation of seismic waves or the interaction of electromagnetic fields with the subsurface.

Example 3.2: Imagine you're a detective trying to locate a hidden vault within a building. You might use blueprints and trigonometry to calculate the precise location based on clues like the angles formed by laser beams reflecting off hidden mirrors embedded in the walls. Similarly, geophysicists utilize **algebra, trigonometry, and calculus** to understand and interpret various geophysical phenomena.

Exercise 3.1:

- Explain how trigonometry is used to determine the distance to an underground object detected in a geophysical survey.

Solution 3.1:

Geophysicists can use trigonometry to determine the distance to an underground object by measuring the angles of arrival of signals from the object at different sensor locations on the surface. By knowing the geometry of the sensor arrangement and applying trigonometric functions, they can calculate the distance to the object.

Exercise 3.2:

Magnetometry and Dip Angle Calculation:

In magnetic surveys, geophysicists measure the Earth's magnetic field strength and its inclination (dip angle) at various locations. Explain how trigonometry can be used to determine the **dip angle** of a magnetic anomaly from the measured field strength and the horizontal component of the field.

Solution 3.2:

1. **Measure the total magnetic field strength (T)** and the **horizontal component of the field (H)** using a magnetometer.
2. Apply the **tangent function** to relate the dip angle (θ) to the ratio of the horizontal and total field strengths:

   $\tan(\theta) = H / T$
3. Solve for the **dip angle (θ)** using the inverse tangent function (arctan) on your calculator:

   $\theta = \arctan(H / T)$

   By employing this trigonometric relationship, geophysicists can interpret the dip angle of magnetic anomalies, which can provide valuable clues about the location, size, and orientation of the underlying magnetic source, potentially indicating the presence of specific minerals.

3.2 Calculus and Differential Equations in Geophysics

Example 3.3: Imagine you are studying the flow of water in a river. You might use calculus to analyze the rate of change of water flow over time. Similarly, geophysicists use calculus and differential equations to model the behavior of various geophysical phenomena involving continuous change, such as the propagation of seismic waves or the flow of heat through the Earth's interior.

Example 3.4: Imagine observing ripples spreading outwards on a pond after throwing a pebble. You might use **differential equations** to model the wave propagation and understand the factors influencing the wave behavior. Similarly, geophysicists utilize differential equations to analyze the propagation of seismic waves through the Earth's interior.

Exercise 3.3:

- Briefly describe the role of differential equations in analyzing the propagation of seismic waves.

Solution 3.3:

Differential equations are used to describe the wave equation, which governs the motion and behavior of seismic waves as they travel through the Earth. By solving the wave equation for different Earth models with varying properties, geophysicists can understand how seismic waves propagate, refract, and reflect, allowing them to interpret seismic data and image the Earth's interior.

Exercise 3.4:

Modeling Seismic Wave Attenuation:

Seismic wave amplitude weakens as it travels through the Earth due to various factors like absorption and scattering. Describe how a **differential equation** can be used to model this **amplitude attenuation** and how it can be helpful in interpreting seismic data.

Solution 3.4:

One example of a differential equation used to model seismic wave attenuation is the **damped wave equation**:

$$\frac{\partial^2 u}{\partial t^2} + 2\alpha \frac{\partial u}{\partial t} + V^2 \nabla^2 u = 0$$

where:

- u(x, y, z, t) represents the wave displacement at a specific location (x, y, z) and time (t)

- α is the attenuation coefficient representing the rate of energy loss
- V is the wave velocity
- ∇^2 is the Laplacian operator

By solving this differential equation for specific initial conditions and analyzing the resulting wave behavior, geophysicists can estimate the attenuation coefficient (α) from real seismic data. Knowing the attenuation allows them to account for the weakening signal and improve the accuracy of interpreting the data, leading to a more precise understanding of the Earth's subsurface structure.

3.3 Computer Programming for Data Processing

Example 3.5: Imagine you are analyzing a large dataset of stock prices. It would be time-consuming and error-prone to calculate trends and patterns manually. Similarly, geophysicists deal with vast amounts of data from various sensors and instruments. Computer programming allows them to automate complex calculations, analyze data efficiently, and visualize results for further interpretation.

Example 3.6: Imagine deciphering a coded message. While you might manually analyze each letter and its position, a computer program can efficiently crack the code using specific algorithms. Similarly, geophysicists utilize **computer programming** to tackle complex data analysis tasks in geophysics.

Exercise 3.5:

- Describe two benefits of using computer programming languages in geophysics.

Solution 3.5:

1. **Efficiency and automation:** Programming allows geophysicists to automate repetitive tasks like data processing and calculations, saving time and minimizing errors compared to manual methods.
2. **Data visualization and analysis:** Programming languages enable the creation of sophisticated software tools for visualizing and analyzing large geophysical datasets. These tools can help identify patterns, trends, and anomalies that might be difficult to detect through manual inspection, leading to deeper insights.

These examples and exercises provide a glimpse into the essential mathematical and computational tools used in geophysics. It's important to remember that a deeper understanding of these concepts requires further study and practice.

Exercise 3.6:

Developing a Script for Gravity Anomaly Analysis:

Imagine you have a dataset containing gravity measurements from a specific region. Describe how you would utilize a Python script to:

1. **Import the data** from a file into your Python environment.
2. **Plot the gravity data** as a function of location (e.g., distance along a profile).
3. **Calculate and plot the derivative** of the gravity data to identify potential anomaly locations.

Solution 3.6:

This script demonstrates how to import, visualize, and analyze gravity data in Python to identify potential anomaly locations.

1. Import Libraries:

Python

```
import numpy as np
import matplotlib.pyplot as plt

# Assuming the data is stored in a CSV file
from pandas import read_csv
```

Use code with caution.

content_copy

2. Load Data:

Python

```
# Replace "data.csv" with the actual filename of your data
data = read_csv("data.csv")

# Extract location and gravity values
```

```
locations = data["Location"]  # Assuming the location column is named "Location"
gravity_values = data["Gravity"]  # Assuming the gravity column is named "Gravity"

# Convert data to NumPy arrays for efficient calculations
locations = np.array(locations)
gravity_values = np.array(gravity_values)
```

Use code with caution.

content_copy

3. Plot Gravity Data:

Python

```
# Create the plot
plt.figure(figsize=(10, 6))
plt.plot(locations, gravity_values, marker='o', linestyle='-', label='Gravity (mGal)')
plt.xlabel('Location')
plt.ylabel('Gravity (mGal)')
plt.title('Gravity Anomaly Profile')
plt.grid(True)
plt.legend()
plt.show()
```

Use code with caution.

content_copy

4. Calculate and Plot Derivative:

Python

```
# Calculate the derivative using the 'np.gradient' function
gravity_derivative = np.gradient(gravity_values)

# Plot the derivative along with the original data
plt.figure(figsize=(10, 6))
plt.plot(locations, gravity_values, marker='o', linestyle='-', label='Gravity (mGal)')
plt.plot(locations[1:], gravity_derivative, marker='s', linestyle='-', label='Gravity Derivative
(mGal/m)')
```

```
plt.xlabel('Location')
plt.ylabel('Value')
plt.title('Gravity Profile and Derivative')
plt.grid(True)
plt.legend()
plt.show()
```

Use code with caution.

content_copy

Explanation:

1. The script imports necessary libraries like numpy for numerical computations and matplotlib.pyplot for plotting. Additionally, pandas is used for efficiently reading data from a CSV file.
2. The data is loaded from the CSV file using pandas.read_csv. The script assumes specific column names for location and gravity values, which might need to be adjusted based on your actual data structure. The data is then converted to NumPy arrays for efficient calculations.
3. The script creates a plot using matplotlib to visualize the gravity data as a function of location (e.g., distance along a profile).
4. The derivative of the gravity data is calculated using the np.gradient function. The script then plots both the original data and the derivative on the same figure. This allows for the visualization of potential anomaly locations, as sharp changes in the derivative often indicate the presence of anomalies.

Note:

This is a basic example, and additional functionalities can be incorporated into the script based on specific analysis needs. For instance, you might want to:

- Implement functions to handle different data formats (e.g., text files).
- Add functionalities for data filtering, noise reduction, and anomaly interpretation.
- Integrate the script with other geophysical data analysis tools for a more comprehensive analysis.

Part 2: Concepts of Data Acquisition and Interpretation

Chapter 4: Geophysical Instrumentation and Field Techniques

Obtaining better understanding of the Earth's subsurface necessitates the use of specialized instruments and meticulous field techniques. This chapter studies the tools and strategies employed in various geophysical methods:

4.1 Gravity Surveys

Objective: Gravity surveys strive to map variations in the Earth's subsurface density by measuring minute fluctuations in the gravitational field. These variations, although subtle, offer valuable insights into the composition and structure of the layers beneath our feet.

Instruments:

- **Gravimeters:** At the heart of gravity surveys lies the **gravimeter** (Figure 4.1). This incredibly sensitive instrument measures the **acceleration due to gravity (g)** with exceptional precision, typically in units of milligals (mGal), which is one-thousandth of a Gal (Galileo).

Figure 4.1: Worden Gravimeter (Smithsonian of 29/11/2024, https://www.si.edu/object/nmah_865075)

The principle behind the gravimeter operation often relies on delicate spring systems or balances that experience a slight deflection due to the gravitational pull of the Earth. By measuring this deflection with high accuracy, geophysicists can determine the local gravitational acceleration at a specific location.

Field Techniques:

1. **Station Measurements:** Gravity surveys typically involve establishing a series of **stations** along a designated survey line or grid. At each station, the gravimeter is carefully leveled and stabilized to ensure accurate measurements. The instrument then records the local value of the acceleration due to gravity.

2. **Data Correction:** The raw data collected from the gravimeter needs meticulous correction before meaningful interpretation can be undertaken. These corrections account for various factors that can influence the measured gravity, including:

- **Instrument drift:** Even the most precise instruments can exhibit slight changes in their calibration over time. Corrections are applied to account for this drift, ensuring consistent measurements throughout the survey.
- **Tidal effects:** The Earth's gravity experiences subtle variations due to the gravitational influences of the Moon and the Sun. These tidal effects need to be mathematically removed from the data to isolate the Earth's static gravitational field.
- **Terrain effects:** The presence of nearby mountains or uneven topography can cause localized variations in the gravitational field. Sophisticated corrections, often involving advanced modeling techniques, are employed to account for these terrain effects and isolate the deeper subsurface features responsible for the observed gravity variations.

3. **Data Interpretation:** Once the raw data has been meticulously corrected, geophysicists analyze the processed data to identify **gravity anomalies**. These anomalies represent deviations from the expected background gravity value, potentially indicating variations in the density of the underlying rock formations. By interpreting these anomalies in conjunction with other geological

information and geophysical data, geophysicists can gain valuable insights into the subsurface structure, potentially revealing the presence of:

- **Geological features:** Variations in rock density can be associated with diverse geological features, such as faults, buried river channels, or salt domes. Identifying these features can aid in understanding the geological history of a region.
- **Mineral deposits:** Certain mineral deposits, like ore bodies, often have densities that differ significantly from the surrounding rocks. By analyzing gravity anomalies, geophysicists can target areas with potential mineral resource potential for further exploration.

Therefore, gravity surveys offer a powerful non-invasive method for exploring the Earth's internal architecture, providing valuable information for a wide range of applications, including resource exploration, geological mapping, and even hazard assessment (Blakely, 1995).

4.2 Magnetic Surveys

Objective: Magnetic surveys aim to map variations in the Earth's subsurface magnetic properties by measuring the **strength and direction** of the Earth's magnetic field at various locations. These variations act as a unique fingerprint, revealing valuable details about the composition and structure of the geological layers.

Instruments:

- **Magnetometers:** The workhorse of magnetic surveys, the **magnetometer**, comes in various forms depending on the specific survey application. These instruments measure the total intensity (strength) and, in some cases, the direction (components) of the Earth's magnetic field with high precision. Common types of magnetometers include:
  - **Proton precession magnetometers:** These instruments measure the precession frequency of protons in a strong magnetic field, which is directly proportional to the total intensity of the Earth's field.
  - **Overhauser magnetometers:** These instruments use polarized nuclei to amplify the weak signal of the Earth's field, allowing for highly precise measurements.
  - **Squid magnetometers (Superconducting QUantum Interference Devices):** Offering exceptional sensitivity, these advanced instruments rely on the principles of superconductivity to detect incredibly small variations in the magnetic field.

The choice of magnetometer depends on factors like the desired accuracy, survey type (ground, airborne, marine), and cost considerations.

Field Techniques:

1. **Station Measurements and Data Correction:** Similar to gravity surveys, magnetic surveys often involve establishing a network of **stations** along survey lines or grids. At each station, the chosen magnetometer records the Earth's magnetic field strength and, if applicable, its direction. The collected raw data then undergoes meticulous correction, similar to gravity surveys, accounting for:

- **Instrument drift:** Consistent calibration checks and corrections are crucial to ensure accurate measurements throughout the survey.
- **Diurnal variations:** The Earth's magnetic field exhibits daily variations due to the influence of the Sun. These variations need to be mathematically removed from the data to isolate the anomalies arising from the subsurface.
- **Datum shift:** Magnetic field measurements are referenced to a specific standard, called the magnetic datum. If data from different surveys or regions are combined, corrections are applied to account for any differences in the datums used.

2. **Data Interpretation:** After applying the necessary corrections, geophysicists interpret the processed data to identify **magnetic anomalies**. These anomalies represent deviations from the expected background magnetic field and can be attributed to variations in the magnetic properties of the underlying rocks. The interpretation process often involves:

- **Separation of induced and remnant magnetization:** The Earth's magnetic field can induce magnetization in certain rocks, and some rocks retain a permanent magnetization (remnant magnetization) even when the external field is removed. Distinguishing between these contributions is crucial for accurate interpretation.
- **Modeling and inversion:** Advanced mathematical techniques like modeling and inversion are employed to create models of the subsurface magnetic property distribution that best explain the observed anomalies. This allows geophysicists to estimate the depth, size, and magnetic susceptibility of potential geological features or mineral deposits.

Aeromagnetic Surveys:

For regional exploration and covering large areas efficiently, **aeromagnetic surveys** are often employed. These surveys utilize magnetometers mounted on aircraft or helicopters, allowing for rapid data acquisition over vast regions. Aeromagnetic surveys are particularly valuable for:

- **Mapping regional geological trends:** By providing a broad overview of magnetic anomalies across a large area, aeromagnetic surveys can aid in identifying potential zones of economic interest or mapping large-scale geological structures.
- **Guiding more detailed ground surveys:** The insights gained from aeromagnetic surveys can be used to strategically plan and focus follow-up ground surveys on specific areas of interest identified in the airborne data.

In conclusion, magnetic surveys, with their diverse instrumentation and data interpretation techniques, offer a powerful tool for understanding the Earth's magnetic behaviour. By analyzing the variations in the Earth's magnetic field, geophysicists can gain valuable insights into the composition, structure, and potential resource potential of the geological layers, paving the way for a deeper understanding of our planet's geological history and resource exploration endeavors (Reynolds, 2011; Keating and Fraser, 2009).

4.3 Electrical and Electromagnetic Methods

Objective: Electrical and electromagnetic (EM) methods utilize the principles of electricity and magnetism to probe the Earth's subsurface. These methods are particularly effective in mapping variations in **electrical conductivity**, a key property that reflects the ability of a material to conduct electrical current. By analyzing the interaction of electrical currents and EM fields with the subsurface, geophysicists can gain valuable insights into the composition, structure, and properties of the geological layers.

Instruments:

The diverse nature of electrical and electromagnetic methods necessitates the use of a variety of specialized instruments, each tailored to a specific technique and application. Some common examples include:

- **Resistivity meters:** These instruments measure the **electrical resistance** of the subsurface, which is inversely proportional to conductivity. Resistivity meters often employ two or four electrodes in different configurations to inject electrical current into the ground and measure the resulting voltage difference.

- **Electromagnetic induction (EM) instruments:** EM instruments operate by transmitting a primary EM field into the ground and measuring the induced secondary EM field generated by the interaction with the subsurface. Different types of EM instruments exist, each utilizing different frequencies and configurations to target specific depths and applications. Examples include:
  - **Time-domain EM (TDEM) instruments:** These instruments emit short electromagnetic pulses and measure the response of the subsurface over time, allowing for investigation of deeper layers.
  - **Frequency-domain EM (FDEM) instruments:** These instruments transmit EM fields at specific frequencies and measure the amplitude and phase shift of the received signal, providing information about conductivity variations at different depths.
- **Ground-penetrating radar (GPR) systems:** GPR utilizes high-frequency electromagnetic waves (typically in the radio or microwave range) to image shallow subsurface features with high resolution. By transmitting short radar pulses and analyzing the reflected signals, GPR can map variations in electrical conductivity and permittivity (a measure of how readily a material allows an electric field to form) in the shallow subsurface, making it valuable for applications like:
  - **Environmental investigations:** GPR can be used to map buried objects, locate potential contaminants, and assess soil properties.
  - **Archaeological studies:** GPR can aid in locating and characterizing buried archaeological features and structures.

Field Techniques:

1. **Resistivity Surveys:** These surveys involve injecting electrical current into the ground using electrodes and measuring the resulting voltage difference at other electrodes. The measured electrical response is then used to calculate the **apparent resistivity** of the subsurface. By employing different electrode configurations and interpreting the apparent resistivity variations, geophysicists can map subsurface conductivity variations and identify potential geological features or environmental hazards.
2. **EM Surveys:** EM surveys encompass a wide range of techniques, each with its specific field procedures. In general, they involve:

- Setting up the chosen EM instrument and configuring it for the specific survey objective and target depth.
- Acquiring data along survey lines or grids by transmitting the primary EM field and recording the resulting secondary field response.
- Applying appropriate data processing and corrections to account for factors like instrument drift and environmental influences.

3. **GPR Surveys:** GPR surveys typically involve:

- Utilizing a GPR antenna to transmit short electromagnetic pulses into the ground.
- Recording the reflected signals from the subsurface using the same antenna or a separate receiver antenna.
- Processing the reflected signals to create a two-dimensional image (radargram) representing the variations in electrical properties within the shallow subsurface (Clark & Clark, 1990).

By employing the diverse suite of electrical and electromagnetic methods and interpreting the acquired data effectively, geophysicists can non-invasively gain valuable insights into the sub-surfaces of the earth, contributing to various fields like:

- **Resource exploration:** Identifying potential mineral deposits based on their characteristic electrical conductivity signatures.
- **Environmental investigations:** Mapping buried contaminants, assessing groundwater quality, and locating potential hazards.
- **Civil engineering applications:** Investigating subsurface conditions for infrastructure development and stability analysis.

In conclusion, electrical and electromagnetic methods offer a powerful and versatile approach to get more knowledge of the Earth's subsurface, providing crucial information for a wide range of applications (Ali, 2023; Clark & Clark, 1990).

4.4 Seismic Exploration Techniques

Objective: Seismic exploration techniques utilize **seismic waves**, elastic waves that travel through the Earth, to image the Earth's internal structure and map subsurface features. These powerful tools provide invaluable insights into the composition, layering, and potential resource potential of the geological world.

Instruments:

1. **Seismic Sources:** Seismic surveys rely on generating controlled seismic waves that travel into the Earth's interior. The choice of source depends on various factors, including:

    - **Depth of investigation:** Deeper targets often require stronger sources to generate sufficient energy for the waves to penetrate the Earth's interior.
    - **Environmental considerations:** In sensitive environments, alternative sources like vibrators or even smaller-scale sources like hammer blows are used to minimize environmental impact.

    Common seismic sources include:

    * Explosives: Traditionally used, explosives offer a strong seismic source but are often limited due to environmental concerns and logistical complexities.

    * Vibrators: These controlled sources generate vibrations by applying a varying force to the ground, offering a more environmentally friendly and controllable alternative to explosives.

    * Seismic hammers: These portable sources use a weight that is repeatedly dropped on a plate, generating seismic waves suitable for shallow investigations.

2. **Geophones:** Once the seismic waves are generated, they travel through the Earth's interior and interact with different layers, causing the ground to vibrate. These ground motions are detected and recorded by **geophones**. These highly sensitive instruments convert the ground motion (vibrations) into electrical signals, capturing the characteristics of the seismic waves. Geophones are typically deployed at various locations along a survey line or grid, forming the receiving network.

Field Techniques:

1. **Seismic Reflection:**

    This widely used technique, illustrated in Figure 4.2, operates by:

- **Generating seismic waves** at the surface using the chosen source.
- **Recording the reflected waves** from various underground layers using geophones placed at specific locations.

As the seismic waves travel downwards, they encounter boundaries between different rock layers with contrasting properties. These boundaries act like mirrors, reflecting a portion of the wave energy back towards the surface. The reflected waves are then detected and recorded by the geophones.

By analyzing the **travel time** and **characteristics** of the reflected waves, geophysicists can:

- **Image the subsurface:** Reflected waves arriving from deeper layers take longer to travel and appear later in the recordings. By analyzing the arrival times of these reflections, geophysicists can estimate the depths of the reflecting layers, building a seismic image of the subsurface layering.
- **Identify geological features:** Different rock layers often have distinct seismic wave reflection properties. By interpreting the characteristics of the reflections, geophysicists can potentially identify specific geological features like faults, salt domes, or even potential hydrocarbon reservoirs (Sheriff and Geldart, 1995).

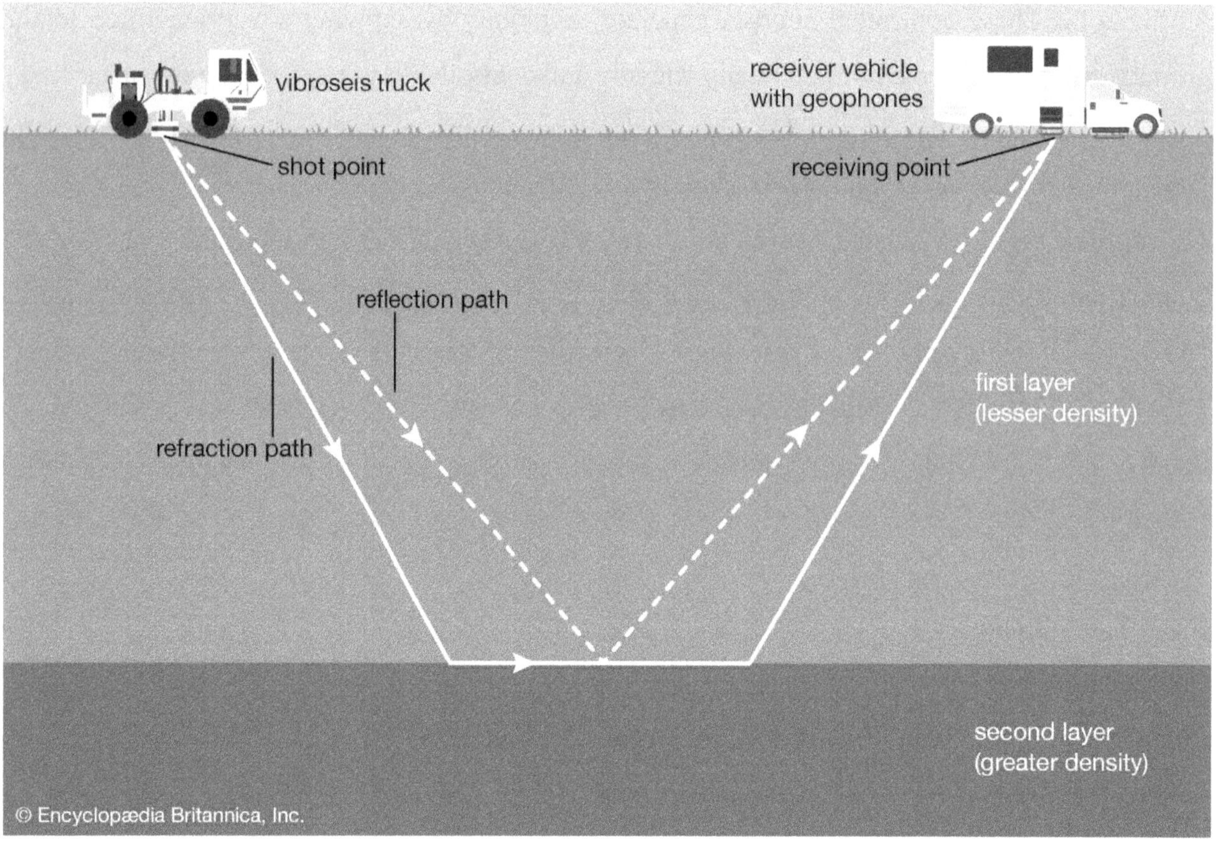

Figure 4.2: Seismic Reflection Survey Setup (Britannica of 29/11/2024; https://cdn.britannica.com/63/4063-050-43F4DCD8/structure-survey-method-natural-gas-petroleum-exploration.jpg)

2. **Seismic Refraction:**

This technique, unlike reflection, focuses on analyzing how seismic waves **travel and bend** through different layers. By measuring the **arrival times** of refracted waves, geophysicists can:

- **Determine layer thicknesses:** The velocities of seismic waves vary depending on the properties of the material they travel through. By analyzing the difference in arrival times of refracted waves traveling through different layers, geophysicists can estimate the thickness and properties of those layers (Dobrin and Savit, 1988).
- **Identify shallow structures:** Seismic refraction is often used for shallow investigations and can be valuable for mapping the thickness of overburden layers or identifying bedrock interfaces.

Choosing the Right Technique:

The selection of the most suitable seismic technique depends on various factors, including:

- **Target property of interest:** Are you searching for specific geological features, investigating near-surface conditions, or aiming to image deeper structures?
- **Depth of investigation:** Different techniques offer varying depths of penetration depending on the source strength and seismic wave characteristics.
- **Logistical considerations:** Factors like environmental restrictions, cost, and accessibility of the survey area can influence the choice of technique.

In many cases, **combining multiple seismic techniques** can provide a more comprehensive understanding of the subsurface. For instance, seismic reflection might be used to identify potential target zones, followed by seismic refraction to determine the thickness and properties of overlying layers in those zones.

Therefore, seismic exploration techniques, with their diverse instrumentation and data interpretation methods, offer a powerful window into the sub-surface layers of the earth, playing a crucial role in resource exploration, geological mapping, and various other applications requiring insights into the Earth's interior structure (Sheriff and Geldart, 1995; Dobrin and Savit, 1988).

Examples and exercises

4.1 Gravity Surveys

Example: Imagine you are searching for a buried treasure chest. You might use a metal detector to identify locations with higher than average metal content, potentially indicating the presence of the chest. Similarly, a gravimeter measures variations in Earth's gravitational pull, which can reveal the presence of denser or less dense materials underground, potentially indicating valuable mineral deposits or geological features.

Exercise 4.1:

Explain why a gravimeter reading might be lower than average over a large underground cave.

Solution 4.1:

A cave represents a large air-filled cavity within the Earth. Air is less dense than the surrounding rock. As a result, the mass present in the area is reduced, leading to a weaker gravitational pull and a lower gravimeter reading compared to the surrounding areas.

Exercise 4.2:

Analyze a real-world gravity data set, available online or through specific software, depicting a specific geological region. Identify any significant gravity anomalies (positive or negative deviations from the average) and discuss their potential geological explanations. Consider factors like the surrounding geology, known geological features, and potential resource implications.

Solution 4.2:

1. **Data Analysis:** Analyze the provided gravity data set, focusing on identifying areas with significant positive or negative anomalies.
2. **Geological Context:** Research and understand the existing geological knowledge of the region, including known rock formations, potential mineral deposits, and any relevant tectonic history.
3. **Anomaly Interpretation:** Based on the anomaly locations and the background geological context, consider:

- **Positive anomalies:** These could indicate denser materials like igneous intrusions, volcanic features, or certain mineral deposits. Correlate the anomalies with existing geological maps and literature to assess their possible origins.

  Negative anomalies: These could suggest less dense features like sedimentary basins, salt domes, or even large underground cavities. Compare them with known geological features and consider potential resource implications, such as groundwater exploration in sediment-filled basins.

4.2 Magnetic Surveys

Example: Imagine you are a treasure hunter using a metal detector. The detector beeps when it encounters metal objects like coins or jewelry. Similarly, a magnetometer measures variations in Earth's magnetic field, which can be influenced by the presence of certain minerals containing magnetic elements like iron, nickel, or cobalt. These variations can help geophysicists locate potential mineral deposits.

Exercise 4.3:

Describe two limitations of using magnetic surveys for archaeological investigations.

Solution 4.3:

1. **Non-magnetic archaeological materials:** Many archaeological artifacts are not made of magnetic materials, such as pottery, wood, or bones. These objects wouldn't be detected by a magnetic survey.
2. **Cultural and environmental disturbances:** Man-made structures like buildings or underground utilities, as well as natural features like volcanic rocks, can also create magnetic anomalies, making it challenging to distinguish them from archaeological features.

Exercise 4.4:

Design a magnetic survey plan for the archaeological site mentioned above. Consider factors like the expected size and depth of potential artifacts, the type of magnetometer to be used, and the layout of the survey grid. Explain how the interpretation of magnetic anomalies could aid in pinpointing areas of archaeological interest.

Solution 4.4:

1. **Survey Design:**
   - **Target size and depth:** Research the types of artifacts expected at the site (e.g., metal tools, weapons, building foundations) and their estimated sizes and depths.
   - **Magnetometer selection:** Choose a magnetometer with appropriate sensitivity and resolution for the target depth and size. Consider using gradiometers to minimize the influence of regional magnetic variations.
   - **Survey grid:** Design a grid layout covering the suspected settlement area with appropriate spacing between measurement points (determined by target size and desired resolution).

2. **Anomaly Interpretation:**
   - **Positive anomalies:** These could indicate the presence of ferrous (iron-containing) artifacts like tools, weapons, or building materials. However, other factors like geological features or man-made structures can also cause positive anomalies.
   - **Negative anomalies:** These might indicate non-magnetic artifacts like pottery, wood, or bones, but they could also be related to natural features like trenches or variations in the local geology.
3. **Integration with other methods:** Combine the magnetic survey results with other archaeological exploration techniques like ground-penetrating radar or electrical resistivity surveys to differentiate between potential cultural features and geological anomalies.

4.3 Electrical and Electromagnetic Methods

Example: Imagine watering your garden. Dry soil offers more resistance to the flow of water compared to wet soil. Similarly, different materials in the Earth's subsurface have varying electrical conductivity. Electrical and electromagnetic methods measure these conductivity variations, which can provide insights into the presence of groundwater, mineral deposits, or environmental contaminants.

Exercise 4.5:

- Briefly explain how ground-penetrating radar (GPR) can be used to locate buried pipes in a construction zone.

Solution 4.5:

GPR transmits high-frequency electromagnetic waves into the ground. These waves reflect off different interfaces between materials with contrasting electrical properties. By analyzing the reflected waves, GPR can detect buried pipes, which typically have different electrical properties compared to the surrounding soil, allowing construction workers to avoid damaging them during excavation.

Exercise 4.6:

Explain the relationship between electrical conductivity and groundwater contamination. How would you interpret the results of an electrical resistivity survey to map the extent of the contamination plume? Discuss any potential limitations or challenges associated with using this method in this specific case.

Solution 4.6:

1. **Conductivity and Contamination:** Contaminated groundwater often contains dissolved salts and other pollutants, increasing its electrical conductivity compared to clean water.
2. **Interpretation:**
   - **Low resistivity zones:** These may indicate areas with higher water conductivity, potentially suggesting the presence of the contamination plume.
   - **High resistivity zones:** These may represent bedrock or less conductive soil/sediment layers.
3. **Limitations:**
   - **Conductivity variations can also be influenced by factors other than contamination, such as natural variations in soil and rock types.**
   - **Clay-rich soils can also exhibit high resistivity, potentially masking the presence of contamination.**

4.4 Seismic Exploration Techniques

Example: Imagine tapping a table and listening to the sound travel and bounce off nearby objects. Similarly, seismic surveys involve generating sound waves (seismic waves) and recording their reflections and refractions from different layers within the Earth. By analyzing these waves, geophysicists can image the Earth's internal structure and locate features like oil and gas reservoirs or geological faults.

Exercise 4.7:

Differentiate between the applications of seismic reflection and seismic refraction methods.

Solution 4.7:

- **Seismic reflection:** This method is primarily used to image the structure of the Earth's subsurface by analyzing the reflections of seismic waves from various layers. This allows geophysicists to identify features like faults, folds, and potential reservoirs for oil and gas exploration.
- **Seismic refraction:** This method focuses on analyzing how seismic waves bend or refract as they travel through layers with varying velocities. It helps determine the thickness and properties of

different layers within the Earth, often used in engineering applications like site characterization for construction projects.

These examples and exercises offer a basic understanding of the diverse tools and techniques employed in geophysical exploration. Remember that real-world applications involve complex procedures and interpretations that require further study and expertise.

Exercise 4.8:

Compare and contrast the data acquisition and interpretation procedures for seismic reflection and seismic refraction surveys. Explain how the results obtained from both methods could be combined to provide a more comprehensive understanding of the subsurface conditions for the bridge construction project.

These examples and exercises offer a more in-depth exploration of the concepts, encouraging critical thinking and applying the knowledge to practical scenarios. Remember, continuous learning and exploring real-world applications are crucial for comprehending the complexities and diverse applications of geophysical methods.

Solution 4.8:

Data Acquisition and Interpretation:

- **Seismic Refraction:**
  - **Acquisition:** Seismic waves are generated at specific points on the survey line, and their arrival times at geophones placed at various distances are recorded.
  - **Interpretation:** By analyzing the differences in arrival times for different geophone locations, the velocities and thicknesses of subsurface layers can be estimated. This allows for depicting the depth and characteristics of the bedrock interface.
- **Seismic Reflection:**
  - **Acquisition:** Similar to refraction, but involves recording reflected waves from deeper interfaces within the Earth.
  - **Interpretation:** By analyzing the travel times and characteristics of the reflected waves, a detailed image of the subsurface layering and potential geological features can be constructed.

Combining Results:

- **Seismic refraction provides information about the depth and properties of the bedrock, crucial for foundation design.**
- **Seismic reflection can reveal deeper geological structures and potential hazards, like faults or cavities, that might influence bridge stability.**

Combining the information from both methods provides a more comprehensive understanding of the subsurface conditions, enhancing the safety and effectiveness of the bridge construction project.

Remember, these are simplified illustrations. Real-world geophysical exploration often involves complex data processing, integration with various techniques, and expertise from experienced geophysicists.

Some Problems

Gravity Surveys

Problem 4.1: Imagine you're a geologist studying a volcanic region. You suspect the presence of an ancient magma chamber, a large underground reservoir of molten rock, beneath the surface. Since magma is denser than the surrounding rock formations, it would exert a slightly stronger gravitational pull. How could you use gravity surveys to support your hypothesis?

Magnetic Surveys:

Problem 4.2: Archaeologists are investigating a suspected ancient settlement site. While traditional excavation methods can be time-consuming and disruptive, a non-invasive approach is desirable. How could magnetic surveys be used to guide their excavation efforts?

Electrical and Electromagnetic Methods:

Problem 4.3: Environmental engineers are investigating a site suspected of groundwater contamination. They need to determine the extent and depth of the contamination plume. How could electrical resistivity surveys be helpful in this scenario?

Seismic Exploration Techniques:

Problem 4.4: A civil engineering team is planning to construct a new bridge across a river. They need to understand the underlying geological structure, particularly the depth and characteristics of the bedrock, to ensure the bridge's stability. How could seismic refraction be employed for this purpose?

Chapter 5: Data Visualization and Interpretation

Understanding information revealed by geophysical surveys requires transforming raw data into meaningful insights. This chapter focuses on the art and science of data visualization and interpretation in geophysics.

5.1 Visualizing Geophysical Data: Images, Maps, Profiles

The raw data collected during geophysical surveys, consisting of numerous numerical values, often requires transformation into a more comprehensible format for meaningful interpretation. This is where the art of **geophysical data visualization** comes into play. By employing various techniques, geophysicists can transform raw data into informative images, maps, and profiles, allowing them to extract valuable insights into the Earth's internal structure and properties.

5.1.1 Images: Revealing Spatial Patterns

Visualizing spatial variations of a specific geophysical parameter is crucial for identifying anomalies and interpreting trends within the data. Common image-based visualization techniques employed in geophysics include:

- **Contour maps:** These maps depict the **spatial distribution** of a geophysical parameter by connecting points with the same value (isopleths) across the survey area. Contour maps offer a quick visual overview of the data, allowing for the identification of **spatial trends** and **anomalies** (areas with significantly higher or lower values compared to the surrounding regions). For instance, a contour map of gravity data might reveal positive anomalies associated with denser subsurface features or negative anomalies indicating less dense structures (see Figure 2.1).
- **Color-coded images:** Similar to contour maps, color-coded images represent the variation of a geophysical parameter using a **color spectrum**. Different colors are assigned to different value ranges, allowing for a more intuitive visual interpretation of spatial variations. These images often provide a clearer distinction of subtle variations compared to contour maps and can be particularly helpful for highlighting specific features of interest.

5.1.2 Maps: Integrating with the Real World

Geophysical data often holds greater significance when viewed in the context of the existing geological setting and infrastructure. This can be achieved by:

- **Overlaying geophysical data on geographical maps:** By superimposing the processed geophysical data (e.g., anomaly maps) onto existing geographical maps (e.g., topographic maps, geological maps), geophysicists can:
  - **Correlate anomalies with known geological features:** This allows for a more informed interpretation of the geophysical data by associating anomalies with specific rock formations, faults, or potential resource deposits.
  - **Plan further investigations:** The spatial context provided by geographical maps can guide the selection of locations for additional data acquisition or drilling campaigns to further investigate identified anomalies.

5.1.3 Profiles: Delving Deeper

While images and maps provide valuable insights into the spatial distribution of geophysical properties, understanding the variations within the Earth's interior requires a different perspective. This is where **geophysical profiles** come into play. These are essentially **cross-sections** of the data, created by plotting the values along a specific line (**vertical profile**) or transect (**horizontal profile**). By analyzing these profiles, geophysicists can:

- **Visualize subsurface variations:** Vertical profiles depict the variations of a geophysical parameter with depth, allowing for the identification of **layering** and potential **boundaries** between different geological formations. This information is crucial for understanding the **subsurface structure** and the potential presence of features of interest, such as mineral deposits or groundwater aquifers.
- **Correlate data from different sources:** Profiles can be used to compare and correlate data acquired using different geophysical methods conducted along the same line or transect. This integrated approach can provide a more comprehensive understanding of the subsurface by combining the strengths of different geophysical techniques.

In conclusion, effective visualization techniques are essential tools in the geophysicist's arsenal, transforming raw numerical data into meaningful images, maps, and profiles. By utilizing these methods, geophysicists can gain valuable insights into the sub-surface layers of the earth, contributing to diverse fields like resource exploration, geological mapping, and environmental investigations.

5.2 Qualitative and Quantitative Interpretation Methods

The journey from raw geophysical data to meaningful insights into the Earth's interior involves a two-pronged approach: **qualitative interpretation** and **quantitative interpretation**. This combined strategy allows geophysicists to extract valuable information about the internal structure of the earth.

5.2.1 Unveiling the Big Picture: Qualitative Interpretation

Qualitative interpretation serves as the **initial stage** of the interpretation process, focusing on identifying patterns, trends, and anomalies within the visualized geophysical data. This stage heavily relies on the **geophysicist's experience and knowledge** in several key areas:

- **Understanding of the specific geophysical method employed:** Each method has its inherent strengths and limitations, influencing the interpretation of the data. Familiarity with the underlying physical principles and expected responses for different geological scenarios is crucial.
- **Knowledge of the local geology:** Geological maps, existing well logs, and previous studies in the region provide valuable context for interpreting anomalies and relating them to known geological features.
- **Visual acuity and pattern recognition skills:** Identifying subtle trends, spatial relationships between anomalies, and potential inconsistencies in the data requires keen observation skills and the ability to recognize patterns.

Through careful analysis of the visualized data (images, maps, and profiles), qualitative interpretation often involves:

- **Identifying anomalies:** Areas with significantly higher or lower values compared to the background are potential indicators of geological features of interest.
- **Recognizing patterns and trends:** Spatial clustering of anomalies, trends in data variations across profiles, or correlations between data from different methods can provide valuable clues about the underlying geological structures.
- **Formulating initial hypotheses:** Based on the observed patterns and anomalies, geophysicists can formulate preliminary interpretations about the potential nature of the subsurface features, guiding further investigation and quantitative analysis.

5.2.2 Unveiling the Numbers: Quantitative Interpretation

Once promising areas are identified through qualitative interpretation, the focus shifts towards **quantitative interpretation**. This stage aims to extract **numerical estimates** of subsurface properties, such as:

- **Density (gravity methods)**
- **Electrical conductivity (electrical and electromagnetic methods)**
- **Seismic velocity (seismic methods)**

Quantitative interpretation primarily relies on two powerful techniques:

- **Forward modeling:** This method involves creating a **mathematical model** of the subsurface with specific assumptions about the geometry, physical properties (density, conductivity, velocity), and thicknesses of the layers. The model is then used to simulate the expected response of the chosen geophysical method for the given subsurface scenario. By iteratively refining the model parameters and comparing the simulated response with the **actual measured data**, geophysicists can gain insights into the potential properties of the subsurface (Menke, 2012).
- **Inversion:** In contrast to forward modeling, which starts with an assumed model and iterates towards a fit with the data, **inversion** techniques begin with the measured data and attempt to **directly estimate** the subsurface properties. This process involves sophisticated mathematical algorithms that seek the model configuration that best reproduces the observed data. Inversion methods often require careful consideration of the inherent limitations of the chosen technique and potential non-uniqueness of solutions (multiple models might explain the data).

By combining the qualitative and quantitative interpretation approaches, geophysicists can transform geophysical data into a wealth of information about the Earth's internal structure and properties. This knowledge plays a crucial role in various applications, including:

- **Resource exploration:** Identifying potential mineral deposits, oil and gas reservoirs, or groundwater resources.
- **Geological mapping:** Mapping subsurface structures like faults, folds, and sedimentary basins.
- **Environmental investigations:** Identifying and characterizing potential contamination plumes or assessing groundwater quality.

It is important to note that both qualitative and quantitative interpretation require careful consideration of various factors, including:

- **Data quality:** The interpretation process is highly dependent on the quality and accuracy of the acquired data.
- **Model assumptions:** Any assumptions made during forward modeling or limitations of the chosen inversion technique need to be acknowledged and considered when interpreting the results.
- **Integration with other sources of information:** Combining geophysical data interpretation with geological knowledge, well log data, and other available information can lead to more robust and reliable conclusions.

In conclusion, effective interpretation of geophysical data requires a blend of qualitative and quantitative approaches. By leveraging both visual inspection and mathematical techniques, geophysicists can better picture layers below the Earth's surface, contributing to a wide range of applications aimed at understanding and managing our planet's resources and environment.

5.3 Integrating Data from Multiple Methods

While individual geophysical methods offer valuable insights, their limitations highlight the importance of **data integration**. By combining information from multiple methods, geophysicists can gain a **more comprehensive and robust understanding** of the subsurface and mitigate the inherent limitations of any single method.

5.3.1 Synergy through Complementary Information:

Each geophysical method offers unique insights into specific **physical properties** of the subsurface. This diversity becomes a powerful tool when data from multiple methods are combined:

- **Complementary nature:** Different methods are often sensitive to different physical properties. For example, gravity methods primarily respond to variations in **density**, while magnetic methods primarily respond to variations in **magnetic susceptibility**. Combining gravity and magnetic data can, therefore, provide valuable complementary information about the subsurface, potentially differentiating between dense but non-magnetic features (e.g., some mineral deposits) and less dense but magnetic features (e.g., volcanic rocks) (Reynolds, 2011).
- **Enhanced understanding:** Integrating data from multiple methods can paint a **more complete picture** of the subsurface by revealing a broader spectrum of information. For instance, combining seismic reflection data, which provides high-resolution imaging of subsurface layers, with

electrical resistivity data, which offers insights into the conductivity and potentially the fluid content of those layers, can lead to a more comprehensive understanding of the subsurface geology and potential resource potential.

5.3.2 Joint Inversion: A Powerful Tool for Integration

Beyond simply analyzing data from different methods side-by-side, advanced techniques allow for even deeper integration through **joint inversion**. This process involves:

- **Simultaneously inverting data from multiple geophysical methods:** Instead of interpreting each dataset independently, joint inversion utilizes sophisticated mathematical algorithms to **combine the information** from all datasets during the model building process.
- **Accounting for inter-method relationships:** While each method provides independent information, there might be inherent relationships between the physical properties they measure (e.g., electrical conductivity and water content in the case of electrical resistivity surveys). Joint inversion algorithms can account for these relationships, potentially leading to **more reliable and robust** subsurface property estimations.

Benefits of Data Integration:

- **Reduced ambiguity:** Combining data from multiple methods can help **reduce ambiguity** in the interpretation, particularly in scenarios where individual methods might provide inconclusive or conflicting results.
- **Improved characterization:** By utilizing a wider range of information, joint inversion can offer a more **refined and detailed characterization** of the subsurface properties, leading to a more accurate understanding of the geological features of interest.
- **Enhanced decision-making:** The comprehensive picture obtained through data integration empowers geophysicists to make **more informed decisions** regarding resource exploration, geological mapping, or environmental investigations.

Challenges of Data Integration:

- **Data quality and compatibility:** Data from different methods need to be of **consistent quality and appropriately scaled** for successful integration.
- **Computational complexity:** Joint inversion often involves computationally **intensive algorithms**, requiring advanced computing resources and expertise in handling complex data sets.

- **Non-uniqueness of solutions:** Even with multiple data sets, the interpretation process might still face inherent **non-uniqueness**, meaning multiple models could potentially explain the combined data. Careful consideration of geological context and additional information is crucial to constrain the interpretation and arrive at robust conclusions.

  In conclusion, integrating data from multiple geophysical methods offers a powerful approach to study the subsurface of the earth. By leveraging the complementary nature of different methods and employing advanced techniques like joint inversion, geophysicists can gain a more comprehensive understanding of the Earth's interior, leading to significant advancements in resource exploration, geological mapping, and various other applications.

5.4 Case Studies: Successful Interpretations in Various Applications

Geophysical data visualization and interpretation empower various scientific and engineering disciplines to understand and manage the Earth's resources and environment. By delving into real-world examples, we can witness the effectiveness of these techniques in various applications:

5.4.1 Mineral Exploration: Unearthing Earth's Resources

Application: Identifying and characterizing potential mineral deposits, essential for extracting valuable resources like copper, gold, and nickel.

Geophysical Methods Employed:

- **Gravity surveys:** Sensitive to density variations, helping identify potentially mineralized zones due to their often higher density compared to surrounding rocks.
- **Magnetic surveys:** Effective for locating certain mineral deposits containing magnetic minerals like iron, nickel, and cobalt.
- **Electrical and electromagnetic surveys:** Measure electrical conductivity variations, aiding in differentiating mineralized zones from surrounding rock formations.
- **Seismic surveys:** Provide high-resolution images of subsurface structures, aiding in understanding the geological context and potential orientation of mineral deposits.

Challenges:

- **Discerning geological noise:** Distinguishing signals related to the target mineral from background geological features or environmental factors can be challenging.

- **Integrating diverse data:** Combining information from multiple methods effectively while acknowledging their inherent limitations requires expertise and careful analysis (Figure 5.1).

Successful Interpretation:

Integrated geophysical surveys, including magnetic, resistivity, and induced polarization (IP) methods, were conducted to investigate the subsurface geology and mineral potential in the first geophysical evaluation of the Neoproterozoic Santa-Quitéria Magmatic Arc (SQMA), known for its iron oxide-apatite (IOA) occurrences. Results revealed resistive and chargeable bodies at varying depths, with shallow IP anomalies correlating with deep magnetic anomalies. Positive IP-chargeability anomalies are associated with alteration zones, shears, and contact zones, while correlations between high chargeability and high/low magnetic values suggest iron formations and metal sulfides, respectively. (Silva et al., 2022)

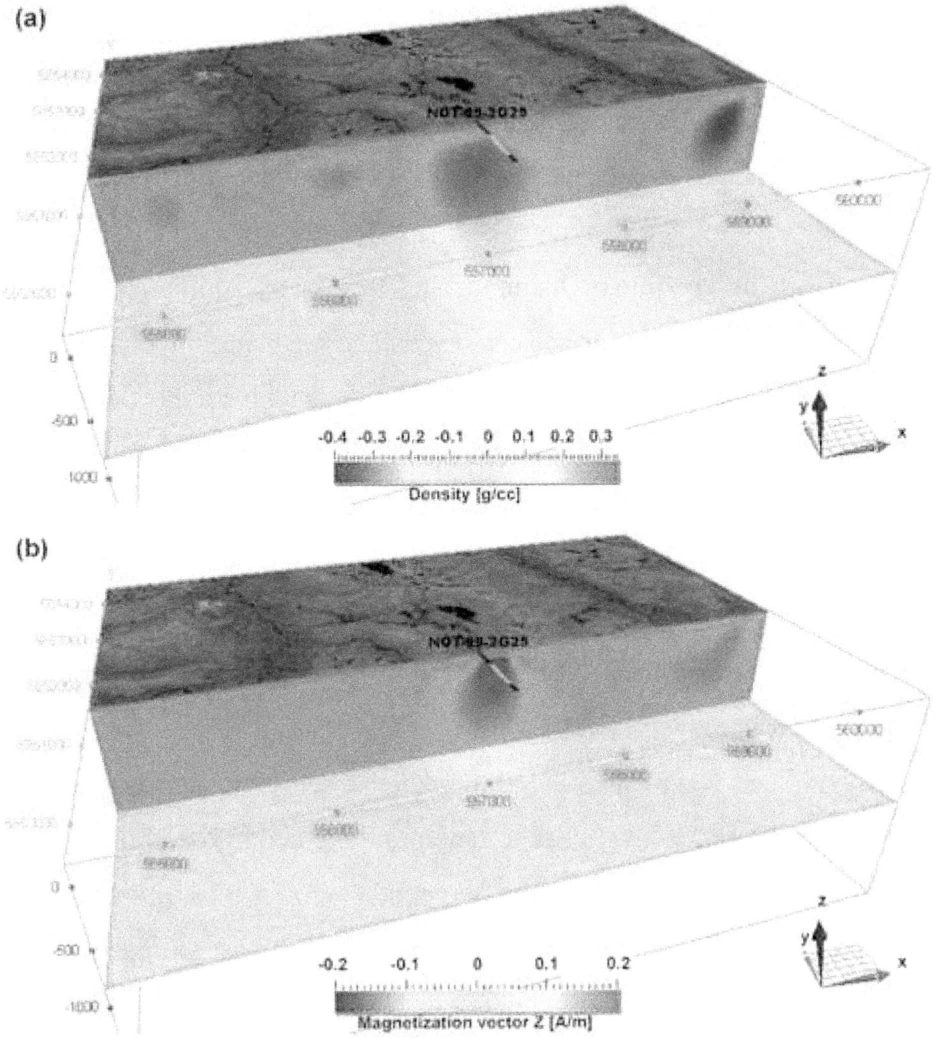

Figure 5.1: Integrated Geophysical Data for Mineral Exploration using gravity and magnetic data (https://pub.mdpi-res.com/minerals/minerals-11-00366/article_deploy/html/images/minerals-11-00366-g012-550.jpg?1628009977 **of 07/03/2024)**

5.4.2 Groundwater Exploration: Mapping the Lifeblood of the Planet

Application: Mapping and characterizing aquifers, crucial for securing sustainable water resources for various human activities.

Geophysical Methods Employed:

- **Seismic refraction surveys:** Measure the velocities of seismic waves to map the depths and thicknesses of different subsurface layers, including potential water-bearing aquifers.

- **Electrical resistivity surveys:** Sensitive to water content and salinity, allowing for the identification and characterization of freshwater aquifers.
- **Ground-penetrating radar (GPR):** Effective in shallow investigations, aiding in mapping the extent and depth of near-surface groundwater resources.

Challenges:

- **Differentiating freshwater from other fluids:** Distinguishing freshwater aquifers from saline water or other conductive fluids within the subsurface can be challenging using some methods.
- **Accounting for complex geological settings:** Interpreting data in areas with complex geology or variable soil conditions requires careful consideration and potentially additional data acquisition techniques.

Successful Interpretation:

Electrical resistivity studies have been effective in identifying potential aquifer zones in the Kalahari Desert by mapping subsurface layers with varying electrical conductivity. Low-resistivity zones often indicate water-saturated sediments, suggesting the presence of aquifers. Seismic refraction studies, on the other hand, have been used to determine the depth to bedrock and the thickness of overlying sediments, which can help identify potential aquifer locations and assess their depth and extent. While specific studies combining both techniques for deep aquifer exploration in the Kalahari Desert might be limited, the combined use of these methods can provide valuable insights into the subsurface geology and groundwater potential of the region. (MDPI. https://www.mdpi.com/2072-4292/14/24/6333).

5.4.3 Geohazard Assessment: Mitigating Potential Disasters

Application: Identifying and characterizing potential geohazards, such as earthquakes, landslides, and volcanic eruptions, to safeguard lives and infrastructure.

Geophysical Methods Employed:

- **Gravity surveys:** Can be used to identify subsurface features like faults, which can be zones of weakness susceptible to earthquakes or landslides.
- **Seismic reflection surveys:** Provide high-resolution images of subsurface structures, aiding in mapping fault locations and assessing their potential activity.

- **Electrical resistivity surveys:** Can be used to map variations in water content within the subsurface, which can be a factor in landslide initiation.
- **Magnetotelluric (MT) surveys:** Measure variations in electrical conductivity within the Earth, providing insights into the deep structure and potential earthquake activity.

Challenges:

- **Predicting the timing and severity of events:** While geophysical data can identify areas prone to geohazards, accurately predicting the timing and severity of these events remains challenging.
- **Integrating diverse disciplines:** Geohazard assessment often involves combining geophysical data with geological and geotechnical information, requiring collaboration between various scientific and engineering disciplines.

Successful Interpretation:

Seismic refraction tomography (SRT), electrical resistivity tomography (ERT), with supporting claims from multichannel analysis of surface waves (MASW) and borehole logs, were used in a geotechnical engineering application to complete a geohazard assessment at a site located adjacent to a river in the mountains of northern British Columbia (Rabczak et al., 2023)

5.4.4 Environmental Monitoring: Safeguarding Our Planet

Application: Detecting and monitoring environmental contamination, such as soil and groundwater pollution, to protect public health and the environment.

Geophysical Methods Employed:

- **Electrical resistivity surveys:** Sensitive to variations in water content and electrical conductivity, aiding in mapping the extent and plume characteristics of contaminant plumes.
- **Electromagnetic conductivity mapping (EM mapping):** Provides high-resolution mapping of near-surface conductivity variations, aiding in identifying potential contamination sources.
- **Ground-penetrating radar (GPR):** Effective in shallow investigations, allowing for the detection of buried waste materials or monitoring the movement of contaminants within the subsurface.

Challenges:

- **Differentiating contamination from other factors:** Distinguishing signals related to the specific contaminant from variations caused by natural geological features or other environmental factors can be challenging.
- **Accounting for complex site conditions:** The effectiveness of geophysical methods for environmental monitoring can be influenced by factors like soil moisture content and the presence of metallic infrastructure.

Successful Interpretation:

electromagnetic low frequency conductivity mapping, electrical resistivity profiles, chargeability profiles and spectral induced polarization (SIP) soundings were successfully applied to an urban site located in the Paris Basin (France) which determined the extension of an organic phase, and of 2 highly mineralized zones that could be linked to biodegraded and/or with pyrite areas in the study area. (Vaudelet et al., 2011).

These real-world examples demonstrate the diverse applications of geophysical data visualization and interpretation in various fields. By effectively translating raw data into meaningful insights, geophysicists play a crucial role in resource exploration, environmental protection, and ensuring the safety and well-being of our planet and its inhabitants.

It is important to note that the specific methods, challenges, and successful interpretations presented in these case studies are just a few examples. The diverse nature of real-world scenarios necessitates a tailored approach to each situation, requiring careful consideration of the specific objectives, geological context, and available resources.

Examples and exercises

5.1 Visualizing Geophysical Data: Images, Maps, Profiles

Example 5.1: Imagine you have a treasure map with an "X" marking the buried treasure. However, the map is just a blank sheet of paper with various numbers and letters scattered across it. It's hard to understand where the treasure is located based on this information alone. Similarly, raw geophysical data is just a collection of numbers and requires visualization to be interpretable. By creating images like contour maps, color-coded plots, and profiles, geophysicists can "see" the internal patterns and anomalies in the data, potentially revealing the location of valuable resources or geological features.

Example 5.2: Imagine a doctor analyzing an X-ray image. While the raw data captured by the X-ray machine consists of numerical values, the doctor visualizes this data as an image to identify potential fractures or abnormalities. Similarly, raw geophysical data is just a collection of numbers, and visualization techniques like **contour maps, color-coded images, and profiles** are crucial for interpreting this data and revealing internal patterns or anomalies. These visualizations allow geophysicists to "see" the data and identify features like:

- **Spatial variations:** How the measured geophysical parameter changes across the survey area.
- **Anomalies:** Areas with significantly higher or lower values compared to the surrounding regions.
- **Trends:** Directional patterns in the data that might indicate underlying geological structures.

Exercise 5.1:

Explain the advantages of using color-coded images to visualize magnetic survey data compared to simply presenting the data as a table of numbers.

Solution 5.1:

A color-coded image allows for a much quicker and more intuitive understanding of the spatial variations in magnetic data compared to a table of numbers. By assigning different colors to different data ranges, geophysicists can easily identify areas with high or low magnetic values, potentially indicating the presence of specific minerals or geological features. This visual representation helps in identifying patterns and anomalies that might be easily overlooked while scrutinizing a table of raw data.

Exercise 5.2:

Explain the advantages of using contour maps to visualize seismic reflection data compared to displaying the individual data points as a scatter plot.

Solution 5.2:

While a scatter plot can show the individual data points representing reflection times at specific locations, it can be challenging to visualize spatial relationships and identify broader trends in the data. Contour maps, on the other hand, offer several advantages:

- **Improved spatial visualization:** By connecting points with the same seismic reflection time, contour maps create lines (isochrons) that represent subsurface boundaries or interfaces reflecting seismic waves. This visual representation allows for a quicker and more intuitive understanding of the overall **shape and continuity** of subsurface geological structures.
- **Trend identification:** Regional trends, such as dipping layers or faults, become readily apparent when visualized through contour lines, aiding in understanding the overall geological setting and potential resource distribution.
- **Data reduction:** Compared to a scatter plot showing all data points, a contour map provides a summarized representation of the data, highlighting key features while reducing visual clutter, making it easier to focus on important trends and patterns.

5.2 Qualitative and Quantitative Interpretation Methods

Example 5.3: Imagine you are looking at a painting. You can appreciate the overall composition, colors, and shapes without needing to know the exact measurements of each brushstroke. Similarly, in the initial stages of interpreting geophysical data, geophysicists use qualitative methods to identify broad patterns, trends, and anomalies in the visualized data. This initial understanding forms the basis for further quantitative analysis.

Example 5.4: Imagine analyzing a fingerprint. While the overall pattern of ridges and loops is readily visible, detailed measurements of individual ridge widths and spacing might not be necessary to identify its uniqueness. Similar to analyzing a fingerprint, **qualitative interpretation** of geophysical data focuses on recognizing **broad patterns and anomalies** in the visualized data without requiring precise numerical information. This initial understanding lays the groundwork for further **quantitative analysis** that involves extracting specific numerical estimates of subsurface properties.

Exercise 5.3:

Describe the role of local geology knowledge in the qualitative interpretation of gravity data.

Solution 5.3:

When interpreting a gravity anomaly map, knowing the types of rocks present in the local geology is crucial. Different rock types have varying densities, which can influence the gravity signal. By considering the local geology, geophysicists can better interpret the observed anomalies. For example, a high gravity anomaly in an area known for dense rock formations might be less surprising than the same anomaly in an area known for low-density sedimentary rocks.

Exercise 5.4:

Describe the role of knowledge about the specific geophysical method employed in the qualitative interpretation of gravity data.

Solution 5.4:

Understanding the underlying principles of the employed geophysical method is crucial for qualitative interpretation of gravity data. Gravity surveys measure variations in the Earth's gravitational field, which are primarily influenced by the density of subsurface materials. By considering the specific method:

- **Geophysicists can anticipate the expected response for different geological features:** For example, dense rock formations like basalts are expected to cause positive gravity anomalies, while less dense features like sedimentary basins might cause negative anomalies.
- **They can better assess the limitations of the method:** Gravity data is sensitive to the total mass of subsurface materials within a certain depth range. Knowing this limitation helps in interpreting anomalies, as they might represent features at varying depths depending on their size and density.
- **They can identify potential sources of artifacts:** Understanding factors like instrument drift or data acquisition inconsistencies can help distinguish true anomalies from artifacts introduced during data collection or processing.

5.3 Integrating Data from Multiple Methods

Example 5.5: Imagine solving a complex puzzle with multiple pieces. Each piece provides a fragment of the whole picture. Similarly, using data from only one geophysical method might not provide a complete understanding of the subsurface. Combining data from multiple methods, like gravity, magnetic, and seismic surveys, allows geophysicists to gain a more comprehensive picture, similar to piecing together the puzzle.

Example 5.6: Imagine solving a Rubik's Cube. While focusing on a single face might provide some information, solving the entire cube requires considering all faces and their interactions. Similarly, relying solely on data from one geophysical method might offer an incomplete picture of the subsurface. Integrating data from multiple methods, such as gravity, magnetic, and seismic surveys, allows geophysicists to gain a more comprehensive and robust understanding of the subsurface and mitigate the inherent limitations of any single method.

Exercise 5.5:

Explain why combining magnetic and seismic data might be beneficial for exploring for mineral deposits.

Solution 5.5:

Magnetic data can be helpful in identifying areas with potentially valuable minerals containing magnetic elements like iron or nickel. However, magnetic anomalies can also be caused by other factors like certain rock formations. Seismic data, on the other hand, can provide information about subsurface layering and structures. Combining these two data sets can help geophysicists differentiate between magnetic anomalies caused by desirable minerals and those caused by other factors, leading to a more targeted and efficient exploration process.

Exercise 5.6:

Explain why combining electrical resistivity (ER) and seismic refraction (SR) data might be beneficial for mapping groundwater resources.

Solution 5.6:

While electrical resistivity (ER) surveys are sensitive to variations in water content and salinity, they might not always distinguish between fresh and saline water. Additionally, they are often limited in depth penetration. Conversely, seismic refraction (SR) surveys can provide information about subsurface layering and depths to bedrock interfaces. Combining these two methods offers:

- **Improved differentiation of water types:** By analyzing both the electrical conductivity (ER) and seismic velocity (SR) of a subsurface zone, geophysicists can gain a more complete picture of its characteristics. For example, a zone with high conductivity and low seismic velocity might be indicative of a freshwater aquifer, while a zone with high conductivity and high seismic velocity

5.4 Case Studies: Successful Interpretations in Various Applications

Case Study 1: Locating Groundwater Resources

Challenge: A community in a drought-stricken region needs to find new sources of freshwater.

Geophysical Methods: Electrical resistivity surveys were conducted to map variations in subsurface electrical conductivity.

Interpretation: Areas with lower resistivity were identified as potential zones with higher water content, indicating potential groundwater aquifers.

Outcome: Successful drilling in these identified zones yielded freshwater sources, alleviating the water scarcity faced by the community.

Case Study 2: Identifying Potential Geohazards

Challenge: A city located near a fault zone needs to assess the risk of future earthquakes.

Geophysical Methods: Combined gravity and seismic reflection surveys were conducted.

Interpretation: The gravity data revealed a subsurface density variation potentially indicating the presence of the fault zone. Seismic reflection data further confirmed the location and orientation of the fault, allowing for estimations of its potential activity.

Outcome: Based on the interpretation, specific areas with higher earthquake risk were identified. This information informed zoning regulations and mitigation strategies to improve public safety in the city.

Case Study 3: Environmental Monitoring and Remediation

Challenge: An industrial site is suspected of soil and groundwater contamination.

Geophysical Methods: Electrical resistivity imaging (ERI) and ground-penetrating radar (GPR) surveys were conducted.

Interpretation: The ERI data identified a zone with low resistivity, potentially indicating a plume of contaminated material. GPR surveys further confirmed the presence of buried waste materials within the subsurface.

Outcome: Based on the interpretation, the extent and location of the contamination were determined. This information guided targeted excavation and remediation efforts, minimizing the environmental impact and protecting public health.

These case studies showcase the diverse applications of geophysical data visualization and interpretation in various fields. By effectively translating raw data into meaningful insights, geophysicists play a vital role in resource exploration, environmental protection, and ensuring the safety and well-being of our planet and its inhabitants.

Part 3: Putting Knowledge to Work - Diverse Applications

Chapter 6: Mapping the Subsurface

Geophysics plays a pivotal role in understanding the layers beneath the Earth's surface, aiding various endeavors like geological mapping, resource exploration, and hazard identification. This chapter clearly illustrates how geophysical methods empower us to "see" and understand the contents of the subsurface of the earth below our feet.

6.1 Imaging and Characterizing Geological Structures

Building upon the foundations of data visualization and interpretation techniques discussed in Chapter 5, geophysical methods empower geologists and engineers to understanding the internal architecture of the Earth, revealing the intricate details of the geological structures that lie beneath our feet. This chapter concentrates on the two key applications of these techniques:

6.1.1 Mapping Geological Formations: Delineating the Subsurface Landscape

By analyzing **variations in physical properties** like density, magnetism, or electrical conductivity, geophysicists can act as cartographers of the unseen, mapping the boundaries and extent of various geological formations. These formations can include:

- **Faults:** Fractures in the Earth's crust where displacement has occurred, often associated with earthquake activity.
- **Folds:** Deformations of rock layers due to tectonic forces, creating distinctive wave-like patterns in the subsurface.
- **Sedimentary layers:** Layers of deposited rock particles that provide valuable information about the geological history and potential resource presence in an area.

Mapping these formations plays a crucial role in various endeavors:

- **Understanding geological history:** The spatial distribution and characteristics of geological formations provide insights into the **geological processes** that shaped a particular region over time. By analyzing these features, geologists can reconstruct past events like tectonic plate movements, volcanic activity, and sedimentary basin formation.
- **Resource exploration:** Identifying and characterizing formations like sedimentary basins, mineral deposits, or potential hydrocarbon reservoirs is crucial for **resource exploration and**

development. Mapping these formations allows geologists to target areas with higher resource potential, guiding further exploration efforts.

Techniques Employed:

- **Seismic reflection surveys:** Utilizing the principles of sound wave reflection, these surveys provide high-resolution images of subsurface structures, allowing for detailed mapping of formations like faults, folds, and sedimentary layers (Yilmaz, 2008).
- **Gravity surveys:** Sensitive to variations in **density**, gravity anomalies can reveal the presence of deep-seated geological features like buried faults, salt domes, or variations in basement rock composition (Reynolds, 2011).
- **Magnetic surveys:** Primarily used for mapping **magnetic susceptibility** variations, these surveys can aid in identifying igneous and metamorphic rocks, potentially associated with certain mineral deposits or structural features (Telford et al., 1990).

6.1.2 Characterizing Subsurface Structures: Decoding the Physical Properties

Beyond mapping the geometry of geological formations, geophysical data offers valuable insights into the **physical properties** of the subsurface materials within these formations. These properties include:

- **Density:** Measured using gravity surveys, density variations can provide information about the type of rock present (e.g., denser rocks like basalts compared to less dense sedimentary rocks). This information can be crucial for understanding the stability of foundations and assessing resource potential.
- **Porosity:** This refers to the void space within a rock or sediment, often estimated through combined analysis of geophysical data (e.g., electrical resistivity) and geological information. Porosity plays a vital role in groundwater storage and fluid flow within the subsurface.
- **Fluid content:** Electrical conductivity and seismic velocity data can be used to infer the presence and type of fluids (e.g., water, oil, gas) within the subsurface pores. This information is crucial for assessing groundwater resources, hydrocarbon exploration, and understanding the behavior of geological hazards like landslides.

Formulae:

While various complex formulae underlie the interpretation of geophysical data, a few fundamental concepts can illustrate the connection between measured parameters and physical properties:

- **Density (ρ):** This is the mass per unit volume of a material and can be expressed as:

$$\rho = \frac{m}{V} \tag{14}$$

where m is the mass and V is the volume.

- **Porosity (φ):** This represents the fraction of void space within a rock or sediment and is calculated as:

$$\varphi = \left(\frac{V_{void}}{V_{total}}\right) \times 100\ \% \tag{15}$$

where V_{void} is the volume of voids and V_{total} is the total volume of the rock.

Understanding these physical properties is essential for various applications:

- **Evaluating foundation stability:** The density and strength of subsurface materials are crucial for assessing the suitability of a site for construction projects.
- **Assessing groundwater potential:** Identifying formations with high porosity and water content is essential for evaluating the potential for sustainable groundwater extraction.
- **Understanding geohazards:** The physical properties of subsurface materials influence the behavior of geohazards like landslides and earthquakes. For instance, analyzing variations in density and water content can be crucial for assessing landslide risk.

Through advancements in data acquisition, processing, and interpretation techniques, geophysicists are continuously refining their ability to **map and characterize geological structures**. This knowledge serves as a powerful tool for geologists, engineers, and environmental scientists, aiding in various endeavors, including:

- **Resource exploration and development:** By identifying and characterizing potential resource deposits like minerals, hydrocarbons, and groundwater, geophysical methods contribute significantly to resource security and economic development.
- **Infrastructure development:** Ensuring the safety and stability of infrastructure projects like dams, bridges, and buildings requires a thorough understanding of the underlying geological structures and their physical properties.
- **Geohazard mitigation:** Characterizing geological formations and their behavior through geophysical methods allows for the identification of areas susceptible to geohazards like landslides and earthquakes, enabling proactive risk mitigation strategies.

- **Environmental monitoring:** Monitoring changes in subsurface properties, such as fluid flow and contaminant migration, helps ensure environmental protection and responsible resource management.

In conclusion, geophysical data visualization and interpretation techniques serve as powerful tools for studying the sub-surface of the earth. By mapping and characterizing geological structures, geophysicists provide valuable insights that contribute to a diverse range of endeavors, from resource exploration and infrastructure development to geohazard mitigation and environmental protection. As technology continues to evolve, the potential of these techniques to contribute to a more sustainable and informed future is certain to grow.

6.2 Identifying Subsurface Hazards

Geophysical surveys play a vital role in safeguarding communities and infrastructure from the often-unforeseen dangers lurking beneath the surface. By **identifying and characterizing potential subsurface hazards**, these surveys empower engineers and scientists to implement proactive measures, mitigating risks and protecting lives. This section explores how various geophysical methods contribute to the assessment of three major categories of subsurface hazards: earthquakes, landslides, and volcanic eruptions.

6.2.1 Earthquake Risk Assessment: Delving into the Fault Lines

Seismic surveys: The cornerstone of earthquake risk assessment, seismic surveys utilize sound waves to image subsurface structures, including **faults**, which are zones of weakness within the Earth's crust where past displacements have occurred and have the potential for future movement. By analyzing **seismic reflection data**, geophysicists can:

- **Map the location, orientation, and depth of faults:** This information is crucial for understanding the potential activity of a fault and its proximity to population centers or critical infrastructure.
- **Estimate fault displacement potential:** Through advanced analysis techniques, seismic data can be used to assess the likelihood and magnitude of potential future earthquake events on a specific fault (Aki and Richards, 2002).

Electrical resistivity (ER) surveys: While seismic surveys excel at imaging faults, ER surveys can provide complementary insights by identifying zones with **weak or saturated materials** within the subsurface. These zones can **amplify ground shaking** during an earthquake, posing a greater threat to structures and human safety. By identifying such zones, ER surveys can help refine earthquake risk assessments and guide mitigation strategies.

Formulae:

One relevant formula employed in earthquake risk assessment is the **Modified Mercalli Intensity (MMI) scale**, a qualitative measure of earthquake ground shaking intensity:

$$MMI = f(a_{max}) \quad (16)$$

where:

- MMI is the Modified Mercalli Intensity
- a_{max} is the maximum ground acceleration (measured in g, where g is the acceleration due to gravity)

This scale, while not providing a direct prediction of earthquake magnitude, helps categorize the intensity of ground shaking experienced at a specific location, which can be crucial for assessing potential damage and guiding emergency response efforts.

6.2.2 Landslide Hazard Assessment: Mapping Instability

Landslides, the movement of a mass of rock, soil, or debris down a slope, pose a significant threat to life and property in mountainous regions and areas with unstable slopes. Geophysical surveys offer valuable tools for:

- **Mapping subsurface features: Seismic refraction and electrical resistivity surveys** can map subsurface features like **water-saturated zones** and **weak layers** within the soil or rock, which can contribute to slope instability and trigger landslides (Jongmans et al., 2009).
- **Monitoring ongoing landslides:** By repeating geophysical surveys over time, geophysicists can monitor changes in subsurface properties, potentially indicating ongoing movement or the risk of reactivation in previously unstable areas.

6.2.3 Volcanic Hazard Assessment: Peering into the Magma Chamber

Volcanic eruptions, while awe-inspiring displays of nature's power, pose a significant threat to nearby communities and infrastructure. Geophysical methods offer crucial insights for:

- **Imaging magma chambers: Electrical conductivity surveys** can detect the presence of **molten rock (magma)** due to its significantly higher electrical conductivity compared to surrounding rock formations. By mapping the location, size, and depth of magma chambers, geophysicists can assess the potential for future eruptions (Gailler et al., 2019).

- **Assessing volcanic activity: Seismic monitoring** allows for the detection of **tremors** associated with volcanic activity, providing early warnings of potential eruptions. Additionally, **gravity surveys** can identify areas of past volcanic activity and potential future eruption zones based on variations in rock density.

Conclusion:

By employing various geophysical techniques, scientists and engineers can identify threats posed by earthquakes, landslides, and volcanic eruptions. Through comprehensive analysis and interpretation of geophysical data, they can contribute significantly to **mitigating risk**, safeguarding communities, and ensuring a safer future for all.

6.3 Case Studies: Geological Mapping, Mineral Exploration, Environmental Site Investigations

Geophysical methods transcend the realm of theoretical concepts, playing a vital role in various real-world applications. This section contains three prominent case studies, showcasing the diverse ways geophysics aids in unveiling the Earth's internal structures and contributing to various endeavors:

6.3.1 Geological Mapping: Illuminating the Subsurface Landscape

Description: Geological mapping aims to create comprehensive representations of the Earth's surface and near-subsurface geology. Traditionally, this has involved field observations, collecting rock samples, and analyzing them in laboratories. However, geophysical surveys offer a powerful **complementary tool** for:

- **Expanding the scope of observations:** By penetrating the Earth's surface, geophysical methods provide insights into the **subsurface geology**, revealing features and structures that might not be accessible through surface observations alone.
- **Creating detailed maps:** Combining data from various geophysical surveys with traditional mapping techniques allows for the creation of more comprehensive and accurate geological maps.

These maps provide valuable information about the distribution and extent of different rock formations, faults, folds, and other geological features.

Challenges:

- **Interpreting complex data:** Analyzing and interpreting geophysical data requires expertise in various geophysical principles, geological knowledge of the specific region, and the ability to differentiate between signals arising from different geological features.
- **Integrating diverse datasets:** Combining data from multiple geophysical methods and geological observations poses a challenge, requiring careful consideration of data compatibility and scaling for effective integration.

Geophysical Methods Employed:

- **Seismic reflection surveys:** These surveys utilize sound waves to image subsurface structures, providing high-resolution information about the geometry and continuity of geological formations (Yilmaz, 2008)

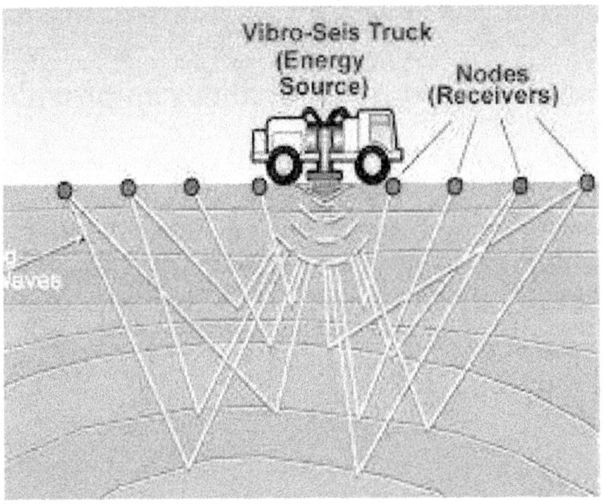

Figure 6.1 Seismic Reflection Survey layout (https://geosiamservices.com/services/ground-services/seismic-survey/seismic-reflection-survey/ on 07/03/2024)

- **Gravity surveys:** Sensitive to variations in **density**, gravity anomalies can reveal the presence of deep-seated geological features like buried faults, salt domes, or variations in basement rock composition (Reynolds, 2011).

Successful Outcomes:

A 3D gravity model was developed in the western Gulf of Mexico in the East Breaks and Alaminos Canyon protraction areas which integrated 3D seismic, gravity and well data and this resulted in a more robust salt model, which has led to significant improvements in subsalt seismic imaging. (Filina et al., 2015).

6.3.2 Mineral Exploration: Unearthing Earth's Treasures

Description: The discovery and extraction of valuable mineral resources are crucial for various industries and play a significant role in global economic development. However, locating these resources can be a challenging and expensive endeavor.

Challenges:

- **Buried deposits:** Valuable mineral deposits often lie buried beneath the Earth's surface, making their direct observation impossible.
- **Vast search areas:** Exploring large areas for potential mineral deposits can be time-consuming and resource-intensive using traditional methods like random drilling.

Geophysical Methods Employed:

- **Gravity surveys:** As different minerals possess varying densities, gravity anomalies can provide clues about the presence of potentially valuable dense mineral deposits like iron or nickel ores.
- **Magnetic surveys:** Minerals with magnetic properties, such as magnetite or hematite, can be detected through magnetic surveys, aiding in identifying potential iron ore, copper, or lead deposits (Telford et al., 1990).

Figure 6.2 Magnetic Survey field experience (www.britannica.com/science/magnetic-survey on 07/03/2024)

- **Electromagnetic (EM) surveys:** Various EM techniques can be employed to detect the electrical conductivity properties of minerals, aiding in the exploration of conductive minerals like copper, nickel, and some sulfide ores.

Successful Outcomes:

A notable example involves the discovery of the giant Grasberg copper and gold deposit in Indonesia. Through a combination of geological mapping, magnetic and gravity surveys, followed by targeted drilling, geologists were able to locate this world-class deposit, contributing significantly to the economic development of the region (IHS Markit, 2020).

6.3.3 Environmental Site Investigations: Safeguarding Our Planet

Description: Environmental site investigations are crucial for assessing potential environmental hazards associated with industrial sites, landfills, or areas suspected of contamination. These investigations aim to:

- **Identify the presence and extent of potential contaminants:** This could include buried waste materials, leaking underground storage tanks, or plumes of contaminated groundwater.
- **Characterize the type and severity of contamination:** Understanding the nature of the contaminants is essential for selecting appropriate remediation strategies.

Challenges:

- **Subsurface heterogeneity:** The subsurface environment can be highly variable, with complex geological formations and varying soil properties, which can complicate the detection and characterization of contaminants.
- **Cost-effective and efficient methods:** Selecting the most appropriate and cost-effective geophysical methods while ensuring they provide the necessary information for accurate site characterization is crucial.

Geophysical Methods Employed:

Electrical resistivity (ER) surveys: By measuring the electrical conductivity of the subsurface, ER surveys can identify zones with **anomalous conductivity** potentially indicative of buried waste materials or contaminated groundwater plumes (Chang et al., 2022).

Figure 6.3 Electrical Resistivity Survey setup (www.geoscan.ca/methods/electrical-resistance on 07/03/2024)

- **Ground-penetrating radar (GPR):** This method utilizes high-frequency electromagnetic waves to image shallow subsurface features with high resolution. GPR surveys are often used to detect buried objects like drums, tanks, or other waste materials.

Figure 6.4 Ground Penetrating Radar Survey field work (https://selectsurveys.com/utility-mapping/gpr-surveys/ on 07/03/2024)

Successful Outcomes:

A successful example involves the identification and characterization of a plume of contaminated groundwater at a former industrial site. By combining ER surveys with groundwater sampling and analysis, environmental engineers were able to map the extent of the contamination plume and identify the type of contaminants present, facilitating the development and implementation of a targeted remediation plan.

Conclusion:

These case studies demonstrate the diverse applications of geophysics in various fields. By better imaging the layers of the earth beneath the surface, geophysical methods contribute significantly to geological mapping, mineral exploration, environmental protection, and ensuring the sustainable development of our planet. As technology continues to advance, the potential of these techniques to contribute to a more informed and responsible future is certain to grow.

Examples and exercises

6.1 Imaging and Characterizing Geological Structures

Example 6.1: Imagine you are building a house. You need to understand the underlying foundation before construction begins. Similarly, geophysicists use various methods to "see" the unseen geological structures beneath the Earth's surface, which are crucial for understanding the stability of the ground and its resource potential. By analyzing variations in physical properties like density, magnetism, or electrical conductivity, they can map the boundaries and characteristics of different geological formations, such as faults, folds, and sedimentary layers.

Example 6.2: Imagine planning a renovation for your home. To ensure structural stability, understanding the underlying foundation is crucial. Similarly, geophysicists use various methods to "see" the unseen geological structures beneath the Earth's surface, essential for understanding its stability and resource potential. By analyzing variations in physical properties like density, magnetism, or electrical conductivity, they can map the boundaries and characteristics of different geological formations, such as faults, folds, and sedimentary layers.

Exercise 6.1:

Explain how a seismic reflection survey can help identify a buried fault in the subsurface.

Solution 6.1:

A seismic reflection survey involves sending sound waves into the Earth and recording their reflections from various layers. When a sound wave encounters a significant boundary between two layers with contrasting properties, like a fault, a portion of the wave reflects back to the surface. By analyzing the reflection patterns and arrival times of the seismic waves, geophysicists can identify the presence, depth, and orientation of buried faults.

Exercise 6.2: Interpreting Fault Displacement using Seismic Reflection Surveys

A seismic reflection survey involves sending sound waves into the Earth and recording the reflected waves returning from various subsurface layers. When these waves encounter a significant boundary between two layers with contrasting properties, like a fault, a portion of the wave reflects back to the surface. By analyzing the reflection patterns and arrival times of the seismic waves, geophysicists can identify the presence, depth, and orientation of buried faults.

Solution 6.2:

1. **Identifying Reflected Signals:** Analyzing the recorded seismic data, geophysicists identify disruptions or offsets in the reflected wave patterns compared to undisturbed areas. These disruptions indicate the presence of a boundary between layers, potentially a fault
2. **Calculating Reflection Depths:** Based on the travel times of the reflected waves and the known speed of sound in different rock types, geophysicists can calculate the depths to the reflecting boundaries. This allows them to estimate the depth of the fault plane.
3. **Determining Fault Orientation:** By analyzing the variations in reflection patterns across the survey area, geophysicists can infer the orientation (dip and strike) of the fault plane. This information is crucial for assessing the potential activity and movement direction of the fault.

Formulae:

- **Travel time (t):** $t = d / v$, where d is the depth to the reflecting boundary and v is the seismic wave velocity in the medium.
- **Reflectivity:** The strength of the reflected signal depends on the acoustic impedance contrast **(Z)** between the two layers, expressed as $Z = \rho v$, where ρ is the density and v is the seismic wave velocity.

6.2 Identifying Subsurface Hazards

Example 6.3: Imagine you are planning a hike in a mountainous region. You would want to be aware of potential hazards like landslides or volcanic activity. Similarly, geophysicists use various methods to identify and assess potential subsurface hazards that could pose a threat to human life and infrastructure.

Example 6.4: While exploring a new hiking trail, understanding potential hazards like landslides or volcanic activity is crucial for safety. Similarly, geophysicists use various methods to identify and assess potential subsurface hazards that could pose a threat to human life and infrastructure.

Exercise 6.3:

- Describe how electrical resistivity surveys can be used to assess the risk of landslides.

Solution 6.3:

Electrical resistivity surveys measure the resistance of the Earth's materials to the flow of electrical current. Water-saturated zones within the subsurface generally have lower resistivity compared to dry rock or soil. By identifying areas with low resistivity through electrical resistivity surveys, geophysicists can potentially locate zones with increased water content and instability, which could be indicative of a higher risk of landslides in specific geological settings.

Exercise 6.4: Assessing Landslide Risk with Electrical Resistivity Surveys

Electrical resistivity (ER) surveys measure the resistance of the Earth's materials to the flow of electrical current. Water-saturated zones within the subsurface generally have lower resistivity compared to dry rock or soil. By identifying areas with low resistivity through ER surveys, geophysicists can potentially locate zones with increased water content and instability, which could be indicative of a higher risk of landslides in specific geological settings.

Solution 6.4:

1. **Mapping Resistivity Variations:** ER surveys involve injecting electrical current into the ground and measuring the resulting voltage differences at various locations. By analyzing the measured resistivity values across the survey area, geophysicists can create a resistivity map depicting variations in subsurface properties.
2. **Identifying Low Resistivity Zones:** Areas with **lower resistivity** on the map potentially indicate the presence of **water-saturated zones** within the subsurface. These zones can contribute to slope instability and trigger landslides, especially when combined with factors like steep slopes and weak rock formations.
3. **Assessing Landslide Risk:** By integrating the ER data with other geological information like slope angle and soil composition, geophysicists can create a **landslide susceptibility map**. This map helps identify areas with a higher risk of landslides, allowing for targeted mitigation strategies like drainage improvements, slope stabilization, or early warning systems.

Formulae: The apparent resistivity (ρ_a) is given by:

$$\rho_a = k \frac{V}{I} \tag{17}$$

where k is a geometrical factor, V is the measured voltage, and I is the injected current.

Factors influencing resistivity: Resistivity depends on various factors including the porosity, water content, and type of pore fluid within the rock or soil.

6.3 Case Studies: Geological Mapping, Mineral Exploration, Environmental Site Investigations

Case Study 1: Mapping a buried Fault for Safe Infrastructure Development

Challenge: A city is planning to build a new bridge across a river. However, concerns exist about the presence of a potentially active fault in the area.

Geophysical Methods: Seismic reflection and gravity surveys were conducted.

Interpretation: Seismic reflection data revealed a clear offset in the rock layers, indicating the presence of a fault. Gravity data provided additional information about the depth and extent of the fault zone.

Outcome: Based on the geophysical data, engineers were able to identify a safer location for the bridge construction, mitigating the risk of damage from future earthquake activity along the fault.

Case Study 2: Mapping a Buried Fault for Safe Infrastructure Development

Challenge: A city is planning to build a new bridge across a river. However, concerns exist about the presence of a potentially active fault in the area.

Geophysical Methods: Seismic reflection and gravity surveys were conducted.

Interpretation: Seismic reflection data revealed a clear offset in the rock layers, indicating the **presence of a fault**. Gravity data provided additional information about the depth and extent of the fault zone, as denser fault zone materials can cause a slight positive gravity anomaly.

Outcome: Based on the geophysical data, engineers were able to identify a safer location for the bridge construction, mitigating the risk of damage from future earthquake activity along the fault. This case study demonstrates how geophysics plays a vital role in ensuring the safety and sustainability of infrastructure development projects.

Case Study 3: Unveiling a buried deposit: Mineral Exploration using Multiple Methods

Challenge: A mining company suspects the presence of a copper deposit in a remote, mountainous region. However, traditional exploration methods proved ineffective due to the complex terrain and lack of surface indications.

Geophysical Methods: A combination of gravity, magnetic, and electromagnetic (EM) surveys were employed.

Interpretation: Gravity data identified a subtle density anomaly potentially indicative of the presence of a mineralized zone. Magnetic data revealed weak magnetic anomalies consistent with the expected signature of copper mineralization. Finally, EM surveys further confirmed the presence of a conductive zone, strengthening the evidence for potential copper deposits.

Outcome: By combining the information from all three geophysical methods, geologists were able to pinpoint a specific target area for drilling. Subsequent drilling confirmed the presence of a significant copper deposit, leading to the development of a new mine, creating jobs, and contributing to the region's economic development.

Case Study 4: Safeguarding Our Planet: Environmental Site Investigation

Challenge: An old industrial site is being redeveloped for residential housing. However, concerns exist about potential soil and groundwater contamination from past industrial activities.

Geophysical Methods: Electrical resistivity (ER) and ground-penetrating radar (GPR) surveys were conducted.

Interpretation: ER data identified **zones with low resistivity**, potentially indicating areas with contaminated soil or groundwater plumes. GPR surveys further revealed the presence of buried objects like drums or tanks, potentially containing hazardous materials.

Outcome: Based on the geophysical data, environmental engineers were able to delineate the extent of the contamination and identify the potential source of the contamination. This information allowed for the development of a targeted remediation plan, ensuring the safety of future residents and protecting the environment from further contamination.

These case studies showcase the diverse applications of geophysics in various real-world scenarios. By employing various geophysical methods and interpreting the data effectively, geophysicists play a crucial role in safeguarding our planet, ensuring the sustainable development of resources, and creating a safer world for future generations.

By employing the diverse arsenal of geophysical tools and interpreting the data effectively, we gain a deeper understanding of the Earth's sub-surface, enabling informed decisions for resource exploration, hazard mitigation, and sustainable development.

Chapter 7: Exploring Earth's Resources (Oil & Gas, Water, Minerals)

The Earth's subsurface holds a treasure trove of resources vital to our society, from the energy sources that power our industries to the minerals that build our infrastructure and the water that sustains life. However, these resources are often out of sight, requiring sophisticated techniques to locate and extract them responsibly. This chapter highlights the fascinating world of resource exploration, highlighting the crucial role of geophysics in this endeavor.

7.1 Resource Exploration: A Geophysical Odyssey

Resource exploration, the process of searching for and identifying valuable natural resources like minerals, hydrocarbons (oil and gas), and groundwater, is a complex and multifaceted endeavor. To navigate this geophysical odyssey, a strategic approach incorporating various geological and geophysical techniques is employed. This section presents the key steps involved in this process:

1. Geological Understanding: Laying the Foundation

The journey begins with a thorough understanding of the regional geology. This involves:

- **Reviewing existing geological maps and literature:** These resources provide valuable insights into the geological history, rock formations, and known mineral deposits within the target area.
- **Consulting with geologists familiar with the region:** Their expertise can contribute valuable insights into the geological setting, potential resource types, and past exploration efforts in the area.
- **Analyzing any available geochemical data:** This data can provide clues about the presence of specific elements or minerals associated with the target resource.

2. Geophysical Surveys: Illuminating the Subsurface

Once a preliminary understanding of the regional geology is established, geophysical surveys become the cornerstone of the exploration process. These surveys, as discussed in Chapter 4, utilize various physical principles to map and characterize the subsurface, providing valuable information that might not be readily available through surface observations alone. Some commonly employed methods in resource exploration include:

- **Seismic reflection surveys:** For oil and gas exploration, these surveys utilize sound waves to image subsurface rock layers and identify potential hydrocarbon reservoirs (Yilmaz, 2008).

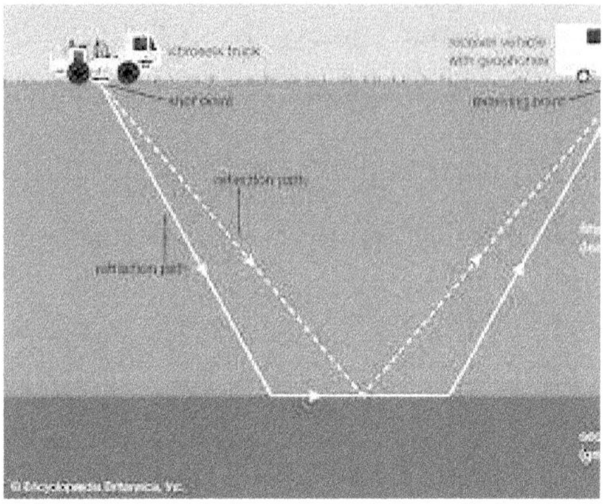

Figure 7.1 Seismic Reflection Survey setup for Oil & Gas Exploration (www.britannica.com/science/seismic-survey on 07/03/2024)

- **Gravity surveys:** Sensitive to variations in density, gravity anomalies can reveal the presence of potentially valuable dense mineral deposits like iron ore or nickel (Reynolds, 2011).
- **Magnetic surveys:** Minerals with magnetic properties, such as magnetite or hematite, can be detected through magnetic surveys, aiding in identifying potential iron ore, copper, or lead deposits (Telford et al., 1990).
- **Electromagnetic (EM) surveys:** Various EM techniques can be employed to detect the electrical conductivity properties of minerals, aiding in the exploration of conductive minerals like copper, nickel, and some sulfide ores.

3. Data Processing and Interpretation: Transforming Raw Data into Insights

The data collected from the geophysical surveys is rarely in a form readily interpretable for resource identification. Therefore, meticulous data processing and interpretation are crucial steps. This stage involves:

- **Data filtering and noise reduction:** Eliminating unwanted noise and enhancing the signal-to-noise ratio for improved data quality.
- **Data correction:** Applying corrections to account for factors like instrument variations or the Earth's natural magnetic field.

- **Data visualization and interpretation:** Utilizing specialized software and expertise, geophysicists transform the processed data into maps and images depicting variations in subsurface properties. These visualizations allow for the identification of potential resource zones, structural features, and anomalies indicative of potential resource presence.

Formulae:

While various complex formulae underlie data processing and interpretation, two fundamental concepts are relevant:

- **Signal-to-noise ratio (SNR):** This ratio measures the strength of the desired signal compared to the background noise and is crucial for assessing data quality. It can be expressed as:

$$SNR = \frac{P_{signal}}{P_{noise}} \tag{18}$$

where:

P_{signal} is the power of the desired signal

P_{noise} is the power of the background noise

- **Apparent resistivity (ρa):** As discussed in Chapter 6, this formula (17) is used in interpreting electrical resistivity surveys and is related to the measured voltage (V), injected current (I), and a geometrical factor (k).

4. Ground-truthing and Well Drilling: Confirming the Discoveries

Once promising targets are identified through the combined efforts of geological understanding and geophysical surveys, **ground-truthing** and **well drilling** become essential for verification. These techniques provide the critical link between the interpreted geophysical data and the actual presence and characteristics of the target resource:

- **Ground-truthing:** This can involve various methods like field mapping, geochemical sampling, or shallow drilling to confirm the presence of any surface expressions or near-surface anomalies identified through the geophysical surveys.
- **Well drilling:** Exploratory drilling, often referred to as well logging, is employed to directly sample the subsurface and obtain physical core samples or analyze the rock formations downhole using specialized logging tools. This allows for the confirmation of the resource type, estimation of its quality and quantity, and evaluation of its economic viability.

By integrating these essential steps, resource exploration ventures can leverage the power of geophysics to navigate the geophysical odyssey, leading to the discovery and development of valuable resources that

7.2 Specific Methods Used for Each Resource

While the core principles of resource exploration outlined in Section 7.1 remain consistent, the specific geophysical methods employed often differ depending on the target resource. This section seeks to explain the nuanced approaches adopted for exploring various resources:

7.2.1 Oil and Gas: Illuminating Subsurface Reservoirs

The search for oil and gas deposits heavily relies on seismic reflection surveys. These surveys, as discussed in Chapter 4 and Section 7.1, utilize sound waves to generate high-resolution images of subsurface rock formations. By analyzing the reflected and refracted seismic waves, geophysicists can:

- **Identify potential hydrocarbon reservoirs:** These reservoirs are often characterized by specific seismic signatures, such as anticlines (upward folds) in the rock layers, which can trap oil and gas (Yilmaz, 2008).
- **Map the geometry and continuity of reservoir formations:** Understanding the size, shape, and connectivity of these formations is crucial for resource assessment and planning for potential extraction.

Formulae:

- **Travel time (t):** As discussed in Chapter 6, this formula remains relevant in interpreting seismic data and relates the depth (d) to a reflecting boundary and the seismic wave velocity (v) in the medium:

$$t = \frac{d}{v} \tag{19}$$

- **Seismic reflection coefficient (R):** This formula quantifies the proportion of the incident seismic wave that is reflected at a boundary between two layers and depends on the acoustic impedance contrast (Z):

$$R = \frac{Z_2 - Z_1}{Z_2 + Z_1} \tag{20}$$

where:

Z_1 and Z_2 are the acoustic impedances of the two layers

In addition to seismic reflection surveys, **gravity and magnetic surveys** can also play a supplementary role in oil and gas exploration:

- **Gravity surveys:** These surveys can aid in identifying **large-scale geological structures**, such as salt domes or regional variations in basement rock composition, which can influence the migration and accumulation of hydrocarbons (Reynolds, 2011).
- **Magnetic surveys:** While not directly detecting hydrocarbons, magnetic surveys can be used to map **basement rocks** and identify structural features that might be associated with potential oil and gas reservoirs.

7.2.2 Groundwater: Mapping the Subsurface Water Flow

The search for groundwater resources often utilizes electrical resistivity (ER) surveys. These surveys, as discussed in Chapter 6, measure the resistance of the Earth's materials to the flow of electrical current. Since water-saturated zones typically have lower resistivity compared to dry rock or soil, ER surveys can be effective in:

- **Mapping variations in water content:** Areas with lower resistivity on the resulting maps may indicate the presence of aquifers or groundwater plumes (Dahlin, 2001).

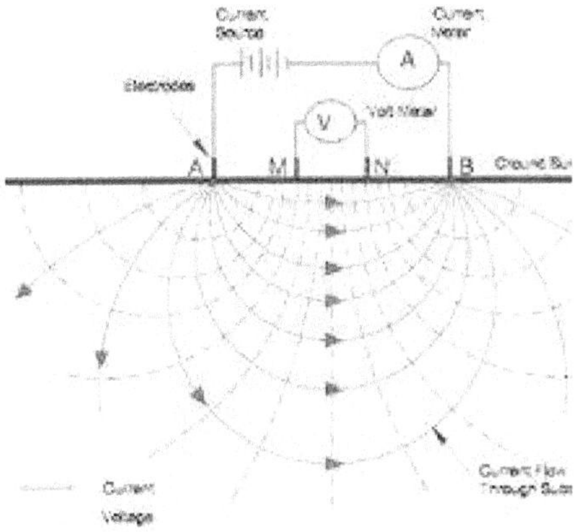

Figure 7.2 Electrical Resistivity Survey setup used for Groundwater Exploration (https://www.researchgate.net/figure/Illustration-of-the-basic-measurements-using-the-method-of-electrical-resistivity_fig3_243464831 on 07/03/2024)

- **Estimating aquifer thickness and depth:** By interpreting the variation in resistivity values with depth, geophysicists can gain insights into the thickness and depth of potential water-bearing zones.

Formula:

Apparent resistivity (ρa): As discussed in Chapter 6, this formula remains relevant for interpreting ER data.

Additionally, seismic refraction surveys can be employed to:

- **Map the boundaries between different water-bearing layers:** By analyzing the refraction of seismic waves at these boundaries, geophysicists can gain insights into the layering and characteristics of the subsurface water system.

7.2.3 Minerals: A Multifaceted Approach

Mineral exploration often involves a **combination of various geophysical methods** due to the diverse properties and characteristics of different mineral deposits. Some commonly employed methods include:

- **Gravity surveys:** As discussed earlier, these surveys can be used to identify dense mineral deposits like iron ore or nickel based on variations in density (Reynolds, 2011).
- **Magnetic surveys:** Minerals with magnetic properties, such as magnetite or hematite, can be directly detected through magnetic surveys, aiding in exploring for iron ore, copper, or lead deposits (Telford et al., 1990).
- **Electromagnetic (EM) surveys:** Various EM techniques, such as induced polarization (IP) surveys, can be used to explore for conductive minerals like copper, nickel, and some sulfide ores based on their distinctive electrical properties. IP surveys measure the chargeability of the subsurface, which refers to the ability of certain materials to store and release electrical energy. This characteristic allows for the differentiation of mineralized zones from barren rock, even when they have similar conductivity (Ward and Barker, 1989).

Formulae:

- **Apparent resistivity (ρa):** As discussed earlier, this formula is relevant for interpreting both ER and IP data

- **Chargeability (M):** Chargeability is often expressed as a percentage and can be calculated using various methods, including:

$$M = (V_s / V_p) * 100\ \% \tag{21}$$

where:

V_s is the secondary voltage (voltage measured after the current is turned off)

V_p is the primary voltage (voltage measured during current injection)

Conclusion:

By employing a strategic combination of geophysical methods tailored to the specific target resource, geophysicists play a crucial role in facilitating the exploration and discovery of valuable resources like oil and gas, groundwater, and minerals. These resources are essential for various aspects of modern life, and responsible exploration practices are crucial for ensuring their sustainable development and minimizing environmental impact.

7.3 Case Studies: Success Stories from the Field

Geophysics transcends the realm of theory, playing a vital role in the real world of resource exploration. This section presents three captivating case studies, showcasing the diverse applications and successes of geophysical methods in identifying and accessing valuable resources:

Case Study 1: Developing Geophysics Applications for Oil and Gas in India (Chopra et al., 2023)

Resource: Oil and Gas

Here is a historical overview of how the Oil and Natural Gas Corporation (ONGC) has utilized geophysics to discover and access oil and gas resources in India:

Early Stages:

- Initially, gravity and magnetic surveys were crucial in identifying broad geological structures conducive to hydrocarbon accumulation. These helped pinpoint areas for further, more detailed investigation.
- These methods helped to establish the initial understanding of India's subsurface geology.

Seismic Revolution:

- The introduction of seismic reflection surveys marked a significant turning point. This method allowed for detailed imaging of subsurface formations, revealing intricate geological structures like faults, folds, and stratigraphic traps.
- 2D seismic surveys were initially used, followed by the adoption of 3D seismic surveys, which provided a more comprehensive and accurate understanding of reservoir geometries.
- Seismic data processing and interpretation techniques evolved over time, improving the ability to identify subtle hydrocarbon indicators.

Well Logging Integration:

- Well logging became an essential tool for evaluating the properties of subsurface formations encountered during drilling.
- Measurements of resistivity, porosity, and density from well logs provided crucial information about the presence and characteristics of hydrocarbon reservoirs.
- This data helped in reservoir characterization and production optimization.

Advanced Techniques:

- ONGC has adopted advanced geophysical techniques, including 4D seismic monitoring, to track changes in reservoirs over time and optimize production.
- Research and development efforts have focused on improving seismic imaging in complex geological settings, such as those found in deep-water areas and beneath volcanic formations.

Role of each method:

- **Gravity and Magnetic:** Provided initial regional surveys to identify potential areas.
- **Seismic:** Provided detailed sub surface imaging to locate precise hydrocarbon deposits.
- **Well Logging:** Provided direct measurements of rock and fluid properties within boreholes.
- **4D Seismic:** Provides monitoring of changes within known reservoirs over time.

Case Study 2: Locating a Sustainable Water Source in the Eastern Sahara (Attwa et al., 2021)

Resource: Groundwater

Challenge:

- Desert environments face severe water scarcity, making groundwater a vital resource.
- Locating and managing groundwater in these regions is challenging due to complex geological structures and limited surface water availability.
- Traditional groundwater exploration methods may be inefficient and unsustainable.

Solution and Methods Used:

- The study proposed an integrated approach combining geophysical and remote sensing methods to enhance groundwater exploration and management.
- **Geophysical Methods:**
    - **Electrical Resistivity Tomography (ERT):**
        - Used to map subsurface resistivity variations, identifying potential aquifer locations based on differences in water content.
        - Helped determine aquifer depths and delineate aquifer boundaries.
    - **Role of ERT:** To identify subsurface zones of low resistivity, which are indicative of water saturated strata.
- **Remote Sensing Methods:**
    - Used to analyze satellite imagery and other remote data to identify geological structures, such as faults and fractures, that control groundwater flow.
    - Helped map watershed boundaries and assess surface water resources.
    - **Role of Remote Sensing:** To identify the geological structures that control the flow of groundwater, and to identify areas where water may accumulate.
- **Integration:**
    - The integration of ERT and remote sensing data provided a comprehensive understanding of the subsurface geology and groundwater potential.
    - This integrated approach improved the accuracy of groundwater exploration and facilitated sustainable water resource management.

Summary:

The research emphasizes the effectiveness of combining geophysical (ERT) and remote sensing methods to improve groundwater exploration in desert environments. By integrating these techniques, researchers were able to more accurately locate potential aquifers and understand the geological structures that control groundwater flow, contributing to more sustainable water resource management.

Case Study 3: Unearthing a Valuable Mineral Deposit in the Haxi mining area of the Xinjiang Uyghur Autonomous Region, China (Xiao-qiang, 2006)

Resource: Gold

Challenge:

- The Haxi gold mining area faced the issue of declining gold reserves due to the exhaustion of easily accessible surface deposits.
- Concealed ore bodies, shielded by overburden or within complex geological structures, were difficult to locate.
- Traditional prospecting methods were proving insufficient for finding these concealed deposits.

Solution and Methods Used:

- The study employed an integrated geophysical approach, combining multiple methods to increase the accuracy and reliability of ore deposit detection.
- **Geophysical Methods:**
    - **Induced Polarization (IP):**
        - Used to detect disseminated sulfide minerals, which are often associated with gold mineralization in the Haxi area.
        - **Role:** Identified zones of high chargeability, indicating the potential presence of sulfide-rich ore bodies.
    - **High-Density Resistivity Method:**
        - Used to map subsurface resistivity variations, which can indicate the presence of altered rock zones associated with gold mineralization.
        - **Role:** Delineated geological structures and identified areas of low resistivity, which may indicate altered rock and potential ore bodies.
    - **Magnetic Surveys:**
        - Used to detect magnetic minerals, which may be associated with gold mineralization or provide information about the geological structure.

        - **Role:** Helped to map geological structures and identify magnetic anomalies that could be related to ore deposits.
    - **Comprehensive interpretation:**
        - The author emphasized the importance of combining the results of these methods and using geological data to improve the accuracy of the final interpretation.

Key Idea:

- The article highlights the effectiveness of integrating multiple geophysical methods to overcome the limitations of individual techniques and improve the detection of concealed ore bodies.
- The combined use of IP, resistivity, and magnetic surveys provided a more comprehensive understanding of the subsurface geology and significantly increased the success rate of gold exploration in the Haxi area.

Summary:

Xiao-qiang, C. demonstrated that by combining multiple geophysical methods, specifically IP, resistivity, and magnetic surveys, and interpreting the results in conjunction with geological data, it is possible to successfully locate concealed ore bodies, thus extending the life of mining operations in areas where surface deposits are depleted.

These case studies showcase the diverse applications of geophysics in resource exploration, contributing to energy security, alleviating water scarcity, and fostering economic development. As technology continues to advance, so too will the capabilities and effectiveness of geophysical methods in responsibly exploring and managing Earth's valuable resources while minimizing environmental impact and ensuring a sustainable future for generations to come.

Examples and exercises

7.1 Resource Exploration: A Geophysical Odyssey

Example 7.1: Imagine you are playing a treasure hunt game. You need to follow clues and solve puzzles to locate the hidden treasure. Similarly, resource exploration involves a series of steps, like understanding the local geology, using different tools (geophysical methods) to gather clues (data) from the hidden resources, and finally verifying the findings through additional techniques like drilling.

Example 7.2: Mapping the Landscape for Mineral Riches

Imagine you're a cartographer tasked with creating a map of a hidden valley rumored to hold valuable mineral deposits. Before venturing into the unknown, you would gather existing maps and information about the surrounding terrain. Similarly, in resource exploration, geologists begin by understanding the regional **geological setting**. This involves:

- **Consulting geological maps and reports:** These resources provide information about the **rock types, their ages, and any known mineral deposits** in the area.
- **Analyzing existing geochemical data:** This data can reveal the presence of **trace elements** associated with specific minerals, acting as potential clues to locate larger deposits.
- **Consulting with geologists familiar with the region:** Their expertise can shed light on past exploration efforts and potential challenges specific to the area.

This initial stage is crucial, as it lays the groundwork for selecting appropriate **geophysical methods** and interpreting the data collected later in the exploration process. Just like a detailed map helps you navigate unfamiliar terrain, understanding the geological context allows geophysicists to embark on a more focused and efficient exploration journey.

Exercise 7.1:

- Explain why understanding the existing geological knowledge is crucial in the initial stage of resource exploration.

Solution 7.1:

Understanding the existing geological knowledge of a region provides valuable context for interpreting geophysical data. Knowing the types of rocks, their distribution, and the geological history of the area allows geophysicists to choose the most appropriate methods and interpret the data effectively. For example, in oil and gas exploration, knowing the presence of specific rock formations known to trap hydrocarbons can guide the selection and interpretation of seismic surveys.

Exercise 7.2:

You are evaluating a region for potential copper deposits. Based on the above information, explain why it's important to consult existing geological maps and reports before conducting any geophysical surveys.

Solution 7.2:

Consulting existing geological maps and reports can be crucial for several reasons:

- **Identifying potential target areas:** Knowing the distribution of different rock types, particularly those known to host copper mineralization, allows you to focus your exploration efforts on areas with higher potential.
- **Choosing appropriate geophysical methods:** Different rock types exhibit varying physical properties. Understanding the geology helps select the most suitable geophysical methods for detecting specific mineral signatures.
- **Interpreting geophysical data effectively:** Existing geological information provides context for interpreting anomalies or variations observed in the geophysical data, leading to more accurate conclusions about potential mineral deposits.

By leveraging existing knowledge, you can significantly increase the efficiency and effectiveness of your exploration endeavors.

7.2 Specific Methods Used for Each Resource

Example 7.3: Imagine searching for different types of objects in a cluttered room. You might use a flashlight to find a lost ring (reflecting light), a magnet to pick up metal objects, and a metal detector for hidden metallic objects. Similarly, different geophysical methods are like specialized tools used to search for specific resources based on their unique physical properties.

Example 7.3: Unmasking the Hidden: Unveiling Mineral Deposits

Imagine searching for various objects buried in a sandbox. A metal detector can easily locate buried metal objects due to their unique magnetic properties. Similarly, geophysical methods exploit the specific physical properties of different resources to differentiate them from the surrounding environment.

Here are some examples of methods used for specific resources:

- **Minerals:** A combination of methods like gravity, magnetic, electromagnetic (EM), and induced polarization (IP) surveys are often employed. Gravity surveys help identify dense mineral deposits like iron ore. Magnetic surveys are effective for minerals with magnetic properties like magnetite or hematite, commonly associated with iron, copper, or lead deposits. EM surveys, particularly IP, can detect conductive minerals like copper, nickel, and some sulfide ores based on their distinctive electrical behavior.

Exercise 7.3:

- Briefly describe how gravity surveys can be helpful in oil and gas exploration, even though oil and gas are not particularly dense materials.

Solution 7.3:

While oil and gas themselves are less dense than surrounding rock formations, they often accumulate in specific geological structures like salt domes or anticlines, which are denser than the surrounding rock. Gravity surveys can map variations in subsurface density, and identifying these denser structures can indirectly indicate potential zones for oil and gas accumulation.

Exercise 7.4:

Explain how electromagnetic (EM) surveys can aid in differentiating between a conductive mineral deposit and a non-conductive one, even if they have similar densities.

Solution 7.4:

While both conductive and non-conductive minerals might exhibit similar densities, their electrical properties differ significantly. EM surveys measure the electrical conductivity of the subsurface. When an EM wave encounters a conductive mineral deposit, it induces electrical currents within the deposit, altering the wave's properties. This change in the EM wave signature can be detected and used to identify the presence of the conductive mineral, even if it has a similar density to the surrounding rock. Conversely, a non-conductive mineral deposit will not significantly affect the EM wave, allowing geophysicists to differentiate between the two based on their contrasting electrical responses.

7.3 Case Studies: Success Stories from the Field

Case Study 1: Finding Water in the Desert

Resource: Groundwater

Challenge: A remote desert community struggles with water scarcity, and finding a sustainable freshwater source is crucial.

Geophysical Methods: Electrical resistivity survey

Outcome: The survey identified a buried paleochannel, an ancient riverbed containing freshwater. Drilling confirmed the presence of a reliable water source, alleviating the water scarcity faced by the community.

Case Study 2: Securing the Future: Unearthing a Geothermal Resource

Resource: Geothermal Energy

Challenge: A region experiences frequent power outages due to an unreliable electricity grid. Identifying a sustainable and renewable energy source is critical for long-term energy security.

Geophysical Methods:

- **Gravity survey:** To map variations in subsurface density, potentially indicating the presence of hot, less dense rock formations associated with geothermal reservoirs.
- **Magnetotelluric (MT) survey:** To measure electrical conductivity variations in the Earth's crust, which can provide insights into the presence and characteristics of geothermal systems.

Outcome:

The combined data from the gravity and MT surveys identified an anomaly indicative of a potential geothermal reservoir. Subsequent drilling confirmed the presence of hot water and steam, paving the way for the development of a geothermal power plant. This project provided a reliable and sustainable energy source for the region, contributing to energy security and environmental sustainability.

This case study demonstrates the growing application of geophysics in exploring and harnessing renewable energy sources like geothermal energy. Additionally, it highlights the potential of geophysics to contribute to a sustainable future by providing clean and reliable energy alternatives.

Case Study 3: From Discovery to Development: Unveiling a Viable Platinum Deposit

Resource: Platinum Group Metals (PGMs)

Challenge: The global demand for platinum, a critical metal used in catalytic converters, is high. Discovering new, economically viable platinum deposits is crucial for meeting this demand.

Geophysical Methods:

- **Magnetic survey:** To detect magnetic anomalies potentially associated with mafic and ultramafic rocks, which are often host rocks for PGMs like platinum.
- **Electromagnetic (EM) survey:** Specifically, an Induced Polarization (IP) survey to identify disseminated sulfide minerals, which can be indicative of PGM mineralization.

Outcome:

The magnetic survey revealed anomalies suggesting the presence of mafic and ultramafic rocks. Subsequent IP surveys identified zones with high chargeability, potentially indicating the presence of sulfide minerals associated with PGMs. Drilling confirmed the presence of a viable platinum deposit. This discovery led to the development of a new mine, creating jobs and contributing to the economic development of the region.

This case study exemplifies the use of complementary geophysical methods for mineral exploration. The magnetic survey provided a broad target area, while the IP survey helped refine the location of potentially valuable PGM mineralization.

These case studies demonstrate the growing application of geophysics in exploring and harnessing renewable energy sources like geothermal energy and in discovering critical mineral deposits like platinum. Additionally, they highlight the potential of geophysics to contribute to a sustainable future by providing clean energy alternatives and responsible resource exploration practices. Remember, these are just a few examples, and geophysics plays a vital role in various resource exploration endeavors beyond these specific cases. As technology continues to evolve, so too will the capabilities and effectiveness of these methods, ensuring their continued importance in responsibly discovering and managing Earth's valuable resources for generations to come.

By employing the power of geophysics, we can embark on a responsible and sustainable journey of resource exploration, ensuring that we meet the needs of the present without compromising the ability of future generations to meet their own.

Part 4: Advanced Topics in Applied Geophysics

Chapter 8: Advanced Seismic Data Processing and Interpretation

This chapter highlights advanced techniques employed in seismic data processing and interpretation, typically used by professionals in the field. These techniques build upon the fundamental concepts discussed in previous chapters and require a solid understanding of seismic theory and signal processing.

8.1 Seismic Wavelet Processing

The **seismic wavelet** plays a crucial role in seismic exploration, acting as the signature of the seismic source and shaping the recorded seismic data. However, the recorded wavelet often differs from the ideal characteristics desired for optimal interpretation. Seismic wavelet processing addresses this issue by employing various techniques to:

8.1.1 Wavelet Estimation: Unveiling the concealed Signature

The first step involves estimating the true wavelet embedded within the recorded seismic data. This is often achieved through deconvolution, a mathematical process that aims to remove the distorting effect of the wavelet, revealing the underlying reflectivity series (Yilmaz, 2008).

There are two main approaches to deconvolution:

- **Deterministic deconvolution:** This method assumes the wavelet is known or can be accurately modeled based on source characteristics and propagation path (Yilmaz, 2008). By applying the inverse of the known wavelet as a filter, the original reflection series can be estimated.
- **Statistical deconvolution:** This approach does not require prior knowledge of the wavelet. Instead, it utilizes statistical relationships within the seismic data to estimate the wavelet and remove its distorting effect (Yilmaz, 2008).

Formulae:

- **Convolution:** The mathematical operation representing the filtering effect of the wavelet on the reflection series (R(t)) can be expressed as:

$$S(t) = W(t) * R(t) \qquad (22)$$

where:

S(t) is the recorded seismic trace

W(t) is the seismic wavelet

R(t) is the reflection series

* denotes the convolution operation

- **Deconvolution:** Aims to find the inverse of the wavelet (W^-1(t)) to recover the reflection series:

$$R(t) = W(t)^{-1} * S(t) \quad (23)$$

8.1.2 Wavelet Shaping: Tailoring the Signal for Interpretation

Once the wavelet has been estimated, wavelet shaping techniques can be applied to modify its characteristics. Ideally, geophysicists strive for a zero-phase wavelet, where the peak of the wavelet coincides with the arrival of the reflected wave, and a narrow peak, which provides better resolution for distinguishing closely spaced reflections (Yilmaz, 2008).

Shaping filters are designed to achieve these desired characteristics by manipulating the frequency content and phase spectrum of the estimated wavelet.

Formulae:

- **Zero-phase wavelet:** A zero-phase wavelet can be obtained by applying a minimum-phase filter, which alters the phase spectrum of the wavelet without affecting its amplitude spectrum (Yilmaz, 2008).

By implementing these processing steps, seismic wavelet processing helps to:

- **Improve the resolution** of the seismic data, allowing for better identification and interpretation of geological features.
- **Enhance the continuity** of seismic reflections, facilitating seismic horizon mapping and interpretation of subsurface structures.
- **Reduce noise** and improve the signal-to-noise ratio, leading to more reliable interpretation of seismic data.

Overall, seismic wavelet processing plays a critical role in transforming the raw seismic data into a form that is more interpretable and informative, ultimately contributing to successful exploration and resource discovery.

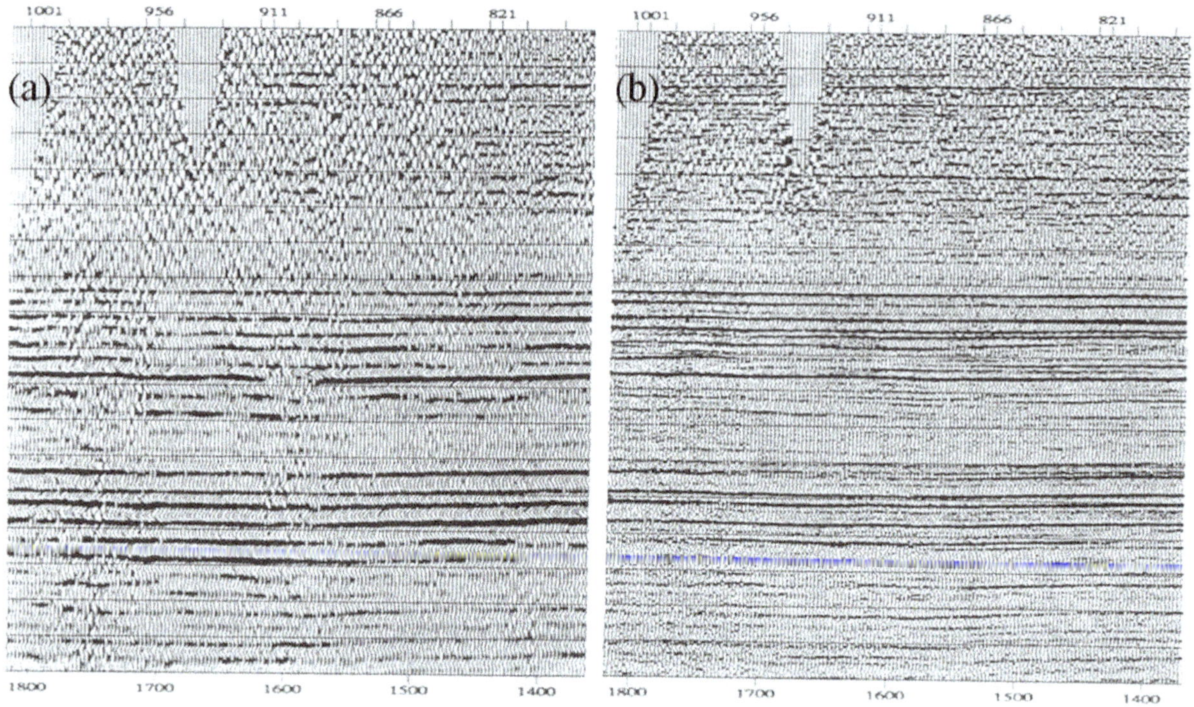

Figure 8.1: Seismic trace before (a) and after deconvolution (b) (Cary 2006)

8.2 Pre-Stack Depth Migration (PSDM)

Seismic waves travel through the Earth at varying velocities depending on the rock properties they encounter. In conventional seismic processing, a fundamental assumption is made: a constant velocity is used to migrate the seismic data, effectively converting travel time from the source to the receiver to depth. However, this assumption often falls short in capturing the complexities of real-world geological settings, where rock properties and velocities can vary significantly. This limitation can lead to distorted and inaccurate images of the subsurface, hindering the interpretation of geological features and potentially hindering successful exploration endeavors.

Pre-stack depth migration (PSDM) addresses this challenge by incorporating varying velocities into the migration process. It utilizes the complete information available in the **pre-stack** seismic data, which includes seismic reflections at various angles of incidence, to achieve a more accurate and detailed image of the subsurface.

Key Advantages of PSDM:

- **Accounting for varying velocities:** Unlike conventional processing, PSDM explicitly considers the variations in seismic velocity within the Earth. This is achieved by incorporating a velocity model into the migration process. This model, often built through geological knowledge, well data, and iterative refinement, provides a map of the expected velocity variations at different depths (Yilmaz, 2008).

- **Improved imaging:** By accounting for these velocity variations, PSDM accurately positions subsurface features in their true depth locations, leading to a more realistic and detailed image compared to conventional migration techniques. This improved imaging allows geophysicists to better interpret complex geological structures like salt domes, fault zones, and dipping beds, which can be challenging to image accurately using simpler migration methods (Alkhalifah et al., 2006).

Formulae:

- **Kirchhoff integral:** This fundamental formula underpins the mathematical framework of PSDM. It describes how seismic wavefronts propagate through a medium with varying velocity, allowing for the calculation of the reflected wavefield at any desired location (Yilmaz, 2008).

$$u(x, y, z, t) = \iint A(x', y', z') \exp(ikT(x', y', z', x, y, z, t)) / T(x', y', z', x, y, z, t)\, dx'\, dy'\, dz' \quad (24)$$

where:

$u(x, y, z, t)$ is the wavefield at a specific point (x, y, z) and time (t)

$A(x', y', z')$ is the amplitude of the wavefront at a source point (x', y', z')

$T(x', y', z', x, y, z, t)$ is the travel time from the source point to the receiver point

k is the wavenumber

- **Velocity model:** The accuracy of PSDM heavily relies on the quality of the velocity model. Common velocity model building techniques include:
  - **Interval velocity estimation:** Estimating the average velocity within specific rock layers based on well data and seismic travel times.
  - **Kirchhoff prestack depth migration (KPSDM):** Using an iterative approach where the initial velocity model is progressively refined based on the results of the migration process itself (Yilmaz, 2008).

In conclusion, PSDM is a powerful tool for seismic imaging, particularly in complex geological settings. By accounting for varying velocities and utilizing the full potential of pre-stack seismic data, PSDM provides a more accurate and detailed representation of the subsurface, enabling geophysicists to make better-informed decisions and ultimately increase the success rate of resource exploration endeavors.

8.3 Reservoir Characterization from Seismic Data

Following the successful identification of potential oil and gas reservoirs through exploration methods, seismic data becomes an invaluable tool for reservoir characterization, a process of understanding the physical properties and characteristics of the reservoir rock that influence its ability to store and produce hydrocarbons. These properties are crucial for determining the economic viability of the reservoir and designing efficient production strategies.

Key Applications of Seismic Data in Reservoir Characterization:

- **Porosity and Permeability Estimation:**
  - **Seismic attributes:** Extracting meaningful information from seismic data often involves calculating seismic attributes. These attributes quantify various characteristics of the seismic signal, such as amplitude, frequency, and continuity, and can be used to estimate reservoir properties like porosity (φ) and permeability (k) (Chaki et al., 2018).
  - **Formulae:** Empirical relationships have been established between specific seismic attributes and reservoir properties based on well log data and core analysis. For example, the following formula relates the acoustic impedance (Z) to porosity:

$$Z = \rho v \tag{25}$$

where:

ρ is the density of the rock

v is the seismic velocity

Changes in porosity can affect the density and, consequently, the acoustic impedance, which can be reflected in certain seismic attributes.

Challenges and limitations: It is crucial to acknowledge that estimating porosity and permeability solely from seismic data is inherently challenging and often requires integration with other data sources like well logs and core samples. Additionally, the relationships between seismic attributes and reservoir properties can be complex and influenced by various factors beyond porosity and permeability, requiring careful calibration and validation.

Fluid Identification:

- **Amplitude Versus Offset (AVO) analysis:** This technique analyzes the variation in seismic reflection amplitude with changing **offset**, the distance between the source and receiver. Different fluids within the reservoir, such as oil, gas, and water, can exhibit distinct reflection characteristics based on their acoustic impedance contrasts with the surrounding rock. By analyzing the AVO response, geophysicists can gain insights into the potential presence and distribution of different fluids within the reservoir (Castagna and Backus, 1993).
- **Formulae:** The Zoeppritz equations provide a theoretical framework for understanding the relationship between reflection coefficient (R), incident angle (θ), and the acoustic properties of the layers involved (Zoeppritz, 1919). AVO analysis utilizes these relationships to interpret the observed variations in reflection amplitude with offset and infer potential fluid presence.

Note: As mentioned, this chapter is intended for readers with a strong background in geophysics and signal processing, as the concepts and formulae discussed require a deeper understanding of these disciplines.

In conclusion, seismic data analysis plays a vital role in reservoir characterization, providing valuable insights into crucial reservoir properties like porosity, permeability, and potentially even fluid content. However, it is crucial to acknowledge the challenges and limitations associated with these interpretations and to integrate seismic data with other sources of information for a more comprehensive understanding of the reservoir.

Examples and exercises

8.1 Seismic Wavelet Processing

Explanation:

Imagine taking a picture with a camera that blurs the image slightly. The blur represents the wavelet's influence on the seismic data. Deconvolution is like trying to "unblur" the image to reveal the original scene more clearly. Similarly, it aims to remove the wavelet's effect from the seismic data, providing a more accurate representation of the subsurface features.

Example 8.1: Imagine recording a sound of a handclap. However, the recording device adds a slight echo to the sound, altering the original signal. This echo acts as the "wavelet" in this case, affecting the recorded data. Deconvolution, similar to using specialized software to remove the echo from the recording, aims to remove the wavelet's influence on the seismic data. This process helps to reveal the true characteristics of the original seismic reflections, leading to a more accurate interpretation of the subsurface geology.

Exercise 8.1: Research and summarize the concept of minimum-phase filtering used in seismic wavelet processing. Explain how it contributes to achieving a desirable wavelet shape.
Solution 8.1: Minimum-phase filtering is a technique used in seismic wavelet processing to achieve a zero-phase wavelet. Unlike regular filters that can introduce phase shifts to the signal, minimum-phase filters only modify the amplitude spectrum of the wavelet without altering its phase spectrum. This ensures that the peak of the wavelet coincides with the arrival of the reflected wave, leading to a sharper and more interpretable signal. By minimizing the phase distortion, minimum-phase filtering contributes to a desirable wavelet shape for improved resolution and accurate identification of geological features in the seismic data.

8.2 Pre-Stack Depth Migration (PSDM)

Explanation:
Imagine driving through a hilly landscape with a map that only shows the roads in a flat projection. This is similar to using conventional seismic migration in complex geological settings. PSDM is akin to having a detailed map that accounts for the hills, providing a more accurate understanding of the actual distances and locations on the road network. This translates to a clearer picture of the subsurface structures when applied to seismic data.
Example 8.2: Imagine navigating through a complex maze using a 2D map that only shows the walls and pathways from above. This is similar to using conventional seismic migration in regions with complex geological features. PSDM, on the other hand, is like having a 3D model of the maze, accounting for the ups and downs of the pathways, providing a more accurate understanding of the maze's structure and allowing for efficient traversal. Similarly, PSDM considers the variations in rock properties and velocities within the Earth, leading to a more realistic and detailed representation of the subsurface structures compared to conventional migration techniques.

Exercise 8.2: Search for online resources or news articles that discuss the application of PSDM in hydrocarbon exploration. Briefly summarize the specific challenges addressed by PSDM in this example and how it contributed to a successful outcome.

Solution 8.2: You can search online for various examples of PSDM used in hydrocarbon exploration. Here's a hypothetical scenario:

An oil and gas company is exploring for hydrocarbons in a region with complex geological structures, including salt domes and dipping beds. Conventional seismic migration techniques struggle to accurately image these features due to the varying velocities within the subsurface. Implementing PSDM allows the company to account for these velocity variations, resulting in a clearer and more accurate image of the subsurface structures. This improved image allows for better identification of potential hydrocarbon reservoirs and facilitates more informed decisions regarding drilling locations, ultimately leading to a successful exploration campaign.

8.3 Reservoir Characterization from Seismic Data

Explanation:

Imagine analyzing the texture and color variations in a photograph of a rock formation to estimate its porosity and the types of minerals present. Similarly, seismic data processing and interpretation techniques extract "attributes" from the data that act like these variations, allowing geophysicists to make informed estimates about the properties of the reservoir rock and the fluids it contains.

Example 8.3: Imagine analyzing a detailed photograph of a tree bark. By observing the texture, color variations, and patterns on the bark, we can gain insights into the tree's age, health, and even the types of insects that might inhabit it. Similarly, by processing and interpreting seismic data, geophysicists can extract various attributes like amplitude, frequency, and continuity, acting as "clues" masked within the data. These attributes can be used to estimate crucial properties of the reservoir rock, such as porosity (the amount of empty space within the rock) and permeability (the ease with which fluids can flow through the rock), ultimately providing insights into the potential presence and distribution of fluids like oil and gas within the reservoir.

Exercise 8.3: Discuss with a partner: How can inaccurate seismic data processing and interpretation potentially impact decisions related to resource exploration and development? What are some potential consequences of such inaccuracies?

Solution 8.3: Inaccurate seismic data processing and interpretation can have significant consequences for resource exploration and development:

- **Misidentification of potential reservoirs:** Inaccurate interpretations might lead to overlooking viable hydrocarbon reservoirs or drilling into unproductive zones, resulting in wasted resources and financial losses.
- **Incorrect well placement:** Inaccurate estimations of reservoir properties can lead to poorly placed wells, potentially missing the target zone or encountering unexpected challenges during drilling, causing delays and increased operational costs.
- **Environmental impact:** Drilling in the wrong locations based on inaccurate seismic data can increase the risk of encountering environmental hazards like sensitive ecosystems or triggering unintended geological events.

These are just some examples, and the specific consequences can vary depending on the context and specific project. Therefore, ensuring accurate and reliable seismic data processing and interpretation is crucial for responsible and successful resource exploration and development.

Other Exercises:

While creating traditional exercises with solutions might not be feasible for this chapter, here are alternative activities that can enhance understanding:

1. **Research and Summarize:** Research a specific technique mentioned in this chapter (e.g., deconvolution, AVO analysis) and prepare a short summary explaining its purpose and applications in simpler terms.
2. **Real-world Applications:** Search for online resources or news articles that discuss the use of advanced seismic data processing techniques in real-world scenarios. Briefly share the key points and discuss the benefits of these techniques in the specific example.
3. **Discussion Prompt:** If you have access to a group of learners with a basic understanding of geophysics, consider facilitating a discussion around the importance of accurate seismic data processing and interpretation in responsible resource exploration and development.

Remember, the goal is to provide a basic understanding of the concepts and their significance even without delving into the complexities of the mathematical and technical aspects.

While these advanced techniques offer deeper insights into the subsurface, they also require specialized software and expertise for proper implementation and interpretation. They play a vital role in the successful exploration and development of valuable resources.

Chapter 9: Geophysical Applications in Environmental and Engineering Studies

Geophysics extends its reach beyond resource exploration, offering valuable tools for environmental and engineering studies. This chapter explores how geophysical methods contribute to various endeavors aimed at protecting our environment and ensuring safe and sustainable development.

9.1 Groundwater Exploration and Monitoring

Groundwater, a vital resource for human consumption, agriculture, and ecosystem health, often lies out of reach beneath the Earth's surface. Geophysical methods play a crucial role in both **groundwater exploration** and **monitoring**, providing valuable information for sustainable water resource management and environmental protection.

9.1.1 Groundwater Exploration: Unveiling Buried Aquifers

Locating potential freshwater aquifers is critical for ensuring sustainable water supplies. Several geophysical methods offer valuable tools for this purpose, each exploiting the contrasting physical properties of different subsurface materials:

- **Electrical Resistivity (ER) Survey:** This method measures the electrical resistance of the subsurface, which varies depending on the water content, salinity, and porosity of the rock formations. Freshwater generally has a higher resistivity compared to saltwater or clay formations. By injecting electrical current into the ground and measuring the resulting potential differences, ER surveys can map variations in resistivity, potentially indicating the presence and depth of freshwater aquifers (Telford et al., 1990).

Formulae:

- **Apparent Resistivity:** The measured resistance during an ER survey is called the apparent resistivity (ρa). It relates to the true resistivity (ρ) of the subsurface and the geometry of the current flow through a complex subsurface using the following formula:

$$\rho_a = K\,\rho \qquad (26)$$

where K is a geometrical factor depending on the electrode configuration used.

- **Interpreting Resistivity Data:** By analyzing the spatial variations of apparent resistivity data, geophysicists can identify zones with higher resistivity, potentially corresponding to freshwater-saturated formations. Advanced inversion techniques can further help estimate the true resistivity distribution and potentially delineate aquifer boundaries.

- **Seismic Refraction Survey:** This method utilizes seismic waves generated by a source (e.g., hammer blow) and recorded by geophones placed at specific distances. The velocity of these waves depends on the density and elastic properties of the subsurface materials. By analyzing the arrival times of seismic waves at different locations, geophysicists can infer the layering and properties of the subsurface, potentially identifying formations with high water content (Reynolds, 2011).
- **Electromagnetic (EM) Surveys:** These methods involve generating a primary electromagnetic field and measuring the secondary field induced in the subsurface. Different types of EM surveys, such as ground-penetrating radar (GPR) and induced polarization (IP), can be used to map variations in water content, conductivity, and clay content, providing insights into the potential presence and characteristics of aquifers (Badora and Wawer, 2018).

9.1.2 Groundwater Monitoring: Tracking Changes and Protecting Resources

Beyond exploration, geophysics also contributes significantly to groundwater monitoring. By employing time-lapse geophysical surveys, repeated measurements are taken over time to track changes in the subsurface properties. This approach offers valuable insights for:

- **Monitoring groundwater levels:** Time-lapse ER surveys, for example, can be used to monitor changes in the resistivity of the subsurface, which can be correlated with variations in groundwater levels, providing crucial information for water resource management and drought monitoring (Chang et al., 2022).
- **Mapping contamination plumes:** Time-lapse EM surveys, particularly IP, can be used to map the migration of contaminants within groundwater, such as saline water intrusion or pollution from industrial activities. This information is essential for environmental protection and remediation efforts (Jin, 1992).

Conclusion:

Geophysical methods play a vital role in both groundwater exploration and monitoring. By providing valuable insights into the presence, distribution, and dynamics of this critical resource, geophysics contributes significantly to sustainable water management practices and environmental protection efforts.

9.2 Geohazards Assessment and Mitigation

Geohazards, including landslides, earthquakes, and volcanic eruptions, pose significant threats to life, infrastructure, and property. Geophysical surveys play a crucial role in assessing and mitigating these risks by providing valuable information about the subsurface and potential geological instabilities.

9.2.1 Landslide Hazard Assessment: Ensuring Slope Stability

Landslides occur when a slope becomes unstable and fails, causing the movement of rock, soil, and debris downhill. Geophysical methods offer valuable tools for assessing landslide risk:

- **Electrical Resistivity (ER) Surveys:** As discussed in section 9.1.1, ER surveys measure the electrical resistance of the subsurface. In the context of landslide assessment, variations in resistivity can be indicative of:
    - **Water saturation:** Zones with higher water content generally exhibit lower resistivity. Identifying areas with significant water saturation within the slope material is crucial, as water can act as a lubricant, reducing friction and increasing the risk of landslides (Telford et al., 1990).
    - **Variations in rock and soil properties:** Different rock and soil types can have varying resistivity values. By analyzing these variations, geophysicists can identify potential weak zones within the slope, contributing to a better understanding of its overall stability.
- **Seismic Refraction Surveys:** Similar to their role in groundwater exploration, seismic refraction surveys can also be used in landslide assessment. By analyzing the seismic velocities of the subsurface layers, geophysicists can:
    - **Map the depth and thickness of different layers:** This information helps identify potential slip surfaces, which are zones of weakness within the slope where failure can initiate (Reynolds, 2011).
    - **Identify areas with potential instability:** Seismic velocities can be influenced by factors like rock density and fracturing. By analyzing variations in seismic velocity, geophysicists can identify zones with potential instability within the slope.

9.2.2 Earthquake Risk Assessment: Preparing for Seismic Events

Earthquakes are sudden movements of the Earth's crust, often triggered by the release of built-up stress along fault lines. Geophysical surveys contribute to earthquake risk assessment by:

- **Seismic Refraction Surveys:** As mentioned earlier, seismic refraction surveys can map the subsurface structures and identify potential fault zones. By understanding the location, depth, and geometry of these faults, scientists can better assess the potential for future earthquakes and develop mitigation strategies.
- **Electrical Resistivity Surveys:** While primarily used for landslide assessment, ER surveys can also provide complementary information for earthquake risk assessment. Identifying zones with **altered rock properties** or **fluid-filled fractures** near identified faults can provide additional insights into potential areas of weakness and seismic activity (Telford et al., 1990).

9.2.3 Volcanic Hazard Assessment: Mitigating Eruption Risks

Volcanic eruptions pose a significant threat to communities located near active volcanoes. **Geophysical surveys** contribute to volcanic hazard assessment through various methods:

- **Electrical Conductivity Surveys:** Measuring the electrical conductivity of the subsurface can help identify the presence of magma chambers, which are molten rock reservoirs beneath the Earth's surface. Monitoring changes in electrical conductivity over time can potentially indicate magma movement and provide early warnings of potential eruptions (Badora and Wawer, 2018).
- **Seismic Surveys:** Seismic waves generated by volcanic activity can be monitored using seismic networks. Analyzing the arrival times and characteristics of these waves can provide information about the location and potential size of an eruption (McNutt et al., 2013).
- **Gravity Surveys:** Gravity anomalies, subtle variations in the Earth's gravitational pull, can be measured using gravimeters. These anomalies can be linked to the presence of dense materials like magma chambers or volcanic deposits. By analyzing gravity data, geophysicists can identify areas of past volcanic activity and potentially predict zones with a higher risk of future eruptions (Reynolds, 2011).

Formulae:

- **Seismic Velocity:** The **seismic velocity (V)** of a wave traveling through a medium is related to the elastic properties of the medium (density, ρ) and the moduli of elasticity (bulk modulus, K; shear modulus, μ) through the following equations:
  - **Compressional wave velocity (Vp):** see equation (5) and (6)
  - **Shear wave velocity (Vs):** See equation (7)

These formulae highlight the importance of understanding the subsurface properties to interpret seismic data and assess potential geohazards.

Conclusion:

Geophysical surveys are invaluable tools for geohazards assessment and mitigation. By providing insights into the subsurface structure, water content, and potential instabilities, geophysics empowers communities to take necessary precautions, develop mitigation strategies, and ultimately safeguard lives, infrastructure, and property from the devastating impacts of geohazards.

9.3 Archaeological Geophysics

Archaeology, the study of past human societies and cultures, often relies on meticulous excavation techniques to uncover and analyze artifacts and features. However, traditional excavation methods can be time-consuming, expensive, and potentially destructive to fragile remains. Archaeological geophysics offers a powerful alternative by utilizing various non-invasive geophysical techniques to explore and map buried archaeological sites without disturbing them.

Key Geophysical Methods in Archaeology:

- **Magnetometry:** This technique measures variations in the Earth's magnetic field caused by buried objects with contrasting magnetic properties compared to the surrounding soil.
    - **Applications:** Magnetometry is particularly effective in detecting:
        - **Ferrous objects:** Iron tools, weapons, and other metallic artifacts can create strong magnetic anomalies, aiding in identifying potential locations of settlements, workshops, or battlefields (Clark and Clark, 1990).
        - **Fired structures:** Bricks, kilns, and other fired structures can also exhibit slight magnetic anomalies compared to natural soils, helping to map building foundations and other architectural features (Clark and Clark, 1990).
    - **Formulae:** The magnetic field strength (B) at a point is measured in units of teslas (T) or nanoteslas (nT). Magnetometers measure the total magnetic field intensity (F), which includes the Earth's main magnetic field (F_0) and any local anomalies caused by buried objects (ΔF):

$$F = F_0 + \Delta F \tag{27}$$

By analyzing spatial variations in the total magnetic field intensity, archaeologists can identify potential locations of buried features with contrasting magnetic properties.

- **Ground Penetrating Radar (GPR):** This method utilizes electromagnetic waves to image the subsurface. A GPR unit transmits high-frequency radio waves into the ground, and the reflected waves are recorded by an antenna. By analyzing the travel time and amplitude of the reflected waves, GPR can create high-resolution images of buried structures and features with varying electrical properties.
  - **Applications:** GPR is well-suited for detecting:
    - **Shallow features:** Walls, foundations, pits, and other relatively shallow archaeological structures can be effectively imaged using GPR due to the limited penetration depth of the high-frequency waves (Conyers and Goodman, 1997).
    - **Graves and burials:** GPR can sometimes help identify potential locations of unmarked graves by detecting variations in soil composition or the presence of associated artifacts (Conyers and Goodman, 1997).
  - **Formulae:** The penetration depth of GPR waves (λ) is influenced by the dielectric permittivity (ε) of the soil:

$$\lambda = \frac{c}{\sqrt{\mu_r + \varepsilon_r}} \tag{28}$$

where:

c is the speed of light in a vacuum

μr is the relative magnetic permeability (usually ≈ 1 for soil)

εr is the relative dielectric permittivity of the soil

Understanding the relationship between permittivity and penetration depth is crucial for interpreting GPR data and determining the optimal survey parameters for specific archaeological applications.

- **Electrical Resistivity Tomography (ERT):** This technique measures the electrical resistance of the subsurface at various depths and locations. Different materials have varying electrical resistivity values, and by injecting electrical current into the ground and measuring the resulting potential differences, ERT can create a map of resistivity variations in the subsurface.
  - **Applications:** ERT is valuable for:

- **Mapping subsurface features:** Variations in resistivity can indicate the presence of buried structures, walls, or ditches due to their contrasting electrical properties compared to the surrounding soil (Ali, 2023).
- **Identifying potential material types:** Different archaeological materials, such as stone foundations, hearths, or filled pits, can exhibit distinct resistivity signatures, aiding in the interpretation of buried features (Ali, 2023).

- **Formulae:** ERT data is often interpreted using Ohm's Law, which relates the current (I) flowing through a conductor to the voltage difference (V) across it and the electrical resistance (R) of the conductor:
- $V = IR$

  By measuring the voltage difference across a specific electrode configuration and injecting a known current, the apparent resistivity of the subsurface can be calculated. Advanced inversion techniques are then used to convert these apparent resistivity measurements into a 3D image (tomogram) of the subsurface resistivity distribution, providing valuable insights into potential archaeological features.

Conclusion:

Archaeological geophysics plays a vital role in modern archaeological investigations by enabling non-invasive exploration and mapping of buried sites. By utilizing various geophysical techniques like magnetometry, GPR, and ERT, archaeologists can gain valuable information about the layout, structure, and potential significance of archaeological sites before committing to potentially disruptive excavation activities. This approach not only preserves fragile archaeological remains but also **optimizes excavation strategies** by focusing on areas with higher potential for significant findings. Additionally, **geophysical data** can be integrated with other archaeological data sources, such as historical records and aerial photographs, to create a more comprehensive understanding of past societies and cultures.

Examples and exercises

9.1 Groundwater Exploration and Monitoring

Example 9.1: Imagine searching for water in a desert. You might use a metal detector to find buried water pipes or look for vegetation that thrives in moist environments. Similarly, geophysicists use various methods like electrical resistivity surveys, which act like advanced tools, to identify potential zones with groundwater reserves.

Example 9.2: Imagine you're planning a camping trip in a remote area with limited access to water sources. While looking for a suitable location, you might notice specific types of vegetation like cacti or desert shrubs, which often thrive in areas with shallow groundwater reserves. Similarly, geophysicists utilize various techniques like electrical resistivity (ER) surveys that act like advanced tools to map variations in subsurface properties, indirectly helping them identify potential zones with freshwater aquifers.

Exercise 9.1:

- Explain why understanding the limitations of geophysical methods is important when searching for groundwater.

Solution 9.1:

While geophysics offers valuable tools for groundwater exploration, it's crucial to understand their limitations. These methods can provide insights into the presence and properties of subsurface materials but cannot directly confirm the presence of water or its quality. Additional techniques like drilling and water sample analysis are often necessary to verify the findings from geophysical surveys.

Exercise 9.2:

Understanding the electrical conductivity of different materials is crucial when interpreting ER data for groundwater exploration. Explain why.

Solution 9.2:

Understanding the electrical conductivity of different materials is critical in interpreting ER data for groundwater exploration because:

- **Freshwater generally has a lower electrical conductivity** compared to saltwater and clay formations. By analyzing the spatial variations in measured resistivity, geophysicists can identify zones with lower conductivity, potentially indicating the presence of freshwater aquifers.
- **Differentiating between different types of rock and soil** is crucial. Each material type possesses a characteristic electrical conductivity. Analyzing these variations helps refine the interpretation of potential aquifer boundaries and avoid misidentification of other conductive materials like clay layers.

  Therefore, understanding the electrical properties of various subsurface materials is essential for accurate interpretation of ER data and successful groundwater exploration.

9.2 Geohazards Assessment and Mitigation

Example 9.3: Imagine living near a steep slope during a heavy rainy season. You might notice cracks appearing in the walls of your house, indicating potential instability. Similarly, geophysicists use electrical resistivity surveys to "see" beneath the surface and identify zones with increased water content or loose materials, which could be indicative of a higher risk of landslides in specific geological settings.

Example 9.4: Imagine living in a coastal region prone to tsunamis. You might be aware of the historical events and evacuation plans in place to safeguard lives and property during such disasters. Similarly, geophysicists utilize seismic refraction surveys to "see" beneath the ocean floor and map potential underwater features, like subduction zones, which play a role in triggering tsunamis and understanding their potential impact on coastal communities.

Exercise 9.3:

- Briefly describe how seismic refraction surveys can contribute to earthquake risk assessment.

Solution 9.3:

Seismic refraction surveys involve sending sound waves into the ground and analyzing their travel times and behavior. By mapping variations in seismic velocities, these surveys can help identify subsurface structures like bedrock and fault zones. Understanding the location and characteristics of these structures is crucial for assessing the potential ground shaking intensity and risk of damage during earthquakes.

Exercise 9.4:

Landslide risk assessment often combines multiple geophysical methods. Explain how combining electrical resistivity (ER) and seismic refraction surveys can provide a more comprehensive picture of potential landslide threats.

Solution 9.4:

Combining ER and seismic refraction surveys can provide a more comprehensive picture of potential landslide threats:

- **ER surveys:** As discussed earlier, ER surveys can identify zones with **higher water content**, which acts as a lubricant and increases the risk of slope instability.
- **Seismic refraction surveys:** These surveys map the **seismic velocities** of subsurface layers, helping identify potential **slip surfaces** within the slope where failure can initiate.

By combining the information from both methods, geophysicists can gain a better understanding of the factors contributing to potential landslide risks. Areas with high water content (identified by ER) and weak zones with low seismic velocity (identified by seismic refraction) can be targeted for further investigation and mitigation strategies.

9.3 Archaeological Geophysics

Example 9.5: Imagine trying to find a buried treasure chest buried in a playground without digging everything up. You might use a metal detector to locate the metal chest. Similarly, archaeologists use magnetometry, which is like a more sophisticated metal detector, to detect buried objects and structures without disturbing the archaeological site.

Example 9.6: Imagine being an art historian trying to locate a lost mural within a historical building. You might use historical records and architectural plans to narrow down potential locations. Similarly, archaeologists utilize ground penetrating radar (GPR), which acts like a sophisticated X-ray, to image buried structures and features within archaeological sites without disturbing the fragile remains.

Exercise 9.5:

Explain why using non-invasive geophysical techniques is crucial in archaeological studies.

Solution 9.5:

Archaeological sites are irreplaceable historical resources. Using non-invasive geophysical techniques like magnetometry or ground penetrating radar allows archaeologists to explore and map these sites without physically disturbing the soil and potentially damaging important artifacts or structures. This helps preserve the integrity of the site for future generations while still gaining valuable knowledge about its history.

Exercise 9.6:

Beyond preserving archaeological sites, non-invasive geophysical techniques offer additional benefits. Briefly describe two advantages of using GPR in archaeological investigations.

Solution 9.6:

GPR offers several advantages in archaeological investigations beyond preserving the site:

1. **High-resolution imaging:** GPR can provide detailed images of **shallow features** like walls, foundations, and even some artifacts, allowing archaeologists to map the layout and potential complexity of the buried structures.
2. **Rapid data acquisition:** Compared to traditional excavation methods, GPR surveys can be conducted relatively quickly, allowing for efficient exploration of large areas and optimizing excavation strategies by focusing on specific zones of interest identified by the GPR data.

These benefits contribute to a more efficient and comprehensive understanding of archaeological sites while minimizing the risk of damaging irreplaceable historical artifacts and structures.

These examples and exercises provide a basic understanding of how geophysics contributes to various environmental and engineering studies.

Geophysical applications in environmental and engineering studies provide valuable tools for safeguarding our planet, ensuring the sustainable use of natural resources, and protecting human life and infrastructure from potential hazards.

References:

- **Aki, K. and Richards, P.G. (2002)** Quantitative Seismology. 2nd Edition, CA: Univ. Sci. Books, Sausalito, 700.
- **Attwa, M., El Bastawesy, M., Ragab, D., Othman, A., Assaggaf, H. M., & Abotalib, A. Z. (2021).** Toward an Integrated and Sustainable Water Resources Management in Structurally-Controlled Watersheds in Desert Environments Using Geophysical and Remote Sensing Methods. Sustainability, 13(7), 4004. https://doi.org/10.3390/su13074004
- **Alkhalifah, T., Al Sharif, S., & Belaid, K. (2006).** Pre-stack 3-D Tau migration and velocity analysis: Application to 3-D seismic data from offshore Abu Dhabi, United Arab Emirates. *GeoArabia, 11*(3). Gulf PetroLink, Bahrain.
- **Blakely, R. J. (1995).** Potential theory in geophysics. Cambridge University Press.
- **Burger, H. R. (1992).** *Exploration geophysics of the shallow subsurface*. Prentice Hall PTR.
- **Cary, P. (2006).** Reflections on the Deconvolution of Land Seismic Data. CSEG Recorder, 31, 25-28
- **Castagna, J. P., & Backus, M. M. (Eds.). (1993).** Offset-dependent reflectivity—Theory and practice of AVO analysis. Society of Exploration Geophysicists. https://doi.org/10.1190/1.9781560802624
- **Chang, P.-Y., Puntu, J. M., Lin, D.-J., Yao, H.-J., Chang, L.-C., Chen, K.-H., Lu, W.-J., Lai, T.-H., & Doyoro, Y. G. (2022).** Using Time-Lapse Resistivity Imaging Methods to Quantitatively Evaluate the Potential of Groundwater Reservoirs. *Water*, *14*(3), 420. https://doi.org/10.3390/w14030420
- **Chen, H., Wang, J., Luo, L., Otto, S., Davis, J., Kuhlman, K. L., & Wu, Y. (2024).** Electrical Resistivity Changes During Heating Experiments Unravel Heterogeneous Thermal-Hydrological-Mechanical Processes in Salt Formations. *Geophysical Research Letters*, *51*(10), e2024GL109836.
- **Chopra, S., Nanda, N. C., & Das, S. K. (2023, November).** Development of geophysics applications for oil and gas exploration in India – a historical ONGC perspective. *GEOHORIZONS*. © SPG India.

- **Clark, O.A., & Clark, A. (1990).** Seeing Beneath the Soil: Prospecting Methods in Archaeology (1st ed.). Routledge. https://doi.org/10.4324/9780203164983
- **Conyers, L.B. and Goodman, D. (1997)** Ground-Penetrating Radar: An Introduction for Archaeologists. AltaMira Press, Walnut.
- Silva, C. D. A., Parente, C. V., Stevanato, R., Canata, R. E., & Oliveira, K. M. L. (2022). Integration of induced polarization, resistivity, and magnetic data of IOA occurrences in Ararenda, western of the Santa Quiteria magmatic arc, Ceará Brazil. *Journal of South American Earth Sciences*, *113*, 103624. https://doi.org/10.1016/j.jsames.2021.103624
- **Dobrin, B.M. and Savit, C.H. (1988) Introduction to Geophysical Prospecting. 4th Edition, McGraw-Hill.**
- **Filina, I., Delebo, N., Mohapatra, G., Coble, C., Harris, G., Layman, J., Strickler, M., & Blangy, J-P. (2015).** Integration of seismic and gravity data — A case study from the western Gulf of Mexico. *Interpretation, 3*, SAC99-SAC106. https://doi.org/10.1190/INT-2015-0050.1
- **Gailler, L., Kauahikaua, J. P., Lénat, J.-F., Revil, A., Gresse, M., Ahmed, A. S., Cluzel, N., Manthilake, G., Gurioli, L., Johnson, T. B., Finizola, A., & Delcher, E. (2019).** 3D electrical conductivity imaging of Halemaʻumaʻu lava lake (Kīlauea volcano). *Journal of Volcanology and Geothermal Research*, *383*, 106527. https://doi.org/10.1016/j.jvolgeores.2019.06.001
- **Chaki, Soumi & Routray, Aurobinda & Mohanty, William. (2018).** Well-Log and Seismic Data Integration for Reservoir Characterization: A Signal Processing and Machine-Learning Perspective. IEEE Signal Processing Magazine. 35. 72-81. 10.1109/MSP.2017.2776602.
- **Jongmans, D., G. Bièvre, S. Schwartz, F. Renalier, N. Beaurez, and Y. Orengo (2009)**, Geophysical investigation of a large landslide in glaciolacustrine clays in the Trièves area (French Alps), *Eng. Geol.*, **109**, 45–56, doi:**10.1016/j.enggeo.2008.10.005**.
- **Keating, P., & Fraser, D. C. (2009).** Electrical and magnetic methods in near-surface geophysics. Society of Exploration Geophysicists.
- **Lowrie, W. (2007).** Geophysics: A very short introduction (Vol. 164). Oxford University Press.
- **Badora, D., & Wawer, R. (2018).** Review of selected geophysical methods used in the assessment of water resources for agriculture. *Polish Journal of Agronomy, 34*, 23–33. https://doi.org/10.26114/pja.iung.328.2018.34.03.

- **Nabighian, M. N. (Ed.). (1988).** *Electromagnetic Methods in Applied Geophysics, Volume 2, Application, Parts A and B*. Society of Exploration Geophysicists
- Ali, M. A. H., Mewafy, F. M., Qian, W., Alshehri, F., Ahmed, M. S., & Saleem, H. A. (2023). Integration of Electrical Resistivity Tomography and Induced Polarization for Characterization and Mapping of (Pb-Zn-Ag) Sulfide Deposits. *Minerals*, *13*(7), 986. https://doi.org/10.3390/min13070986

- **Mcnutt, S. & Thompson, Glenn & West, Michael & Fee, David & Stihler, S. & Clark, Edward. (2013).** Local seismic and infrasound observations of the 2009 explosive eruptions of Redoubt Volcano, Alaska. Journal of Volcanology and Geothermal Research. 259. 1-14. 10.1016/j.jvolgeores.2013.03.016.
- **Menke, W. (2012)** Geophysical Data Analysis: Discrete Inverse Theory. MATLAB Edition, Elsevier, Amsterdam.
- **Ndikum, E. N.**, Koumetio, F., Kenfack J. V. and Tabod, C. T. (2019). Gravity study of the Douala sub-basin (Cameroon) using Euler 3D deconvolution, source edge detection (SED) and special function analysis. SN Applied Sciences https://doi.org/10.1007/s42452-019-1176-y
- **Parasnis, D.S. (1986) Principles of Applied Geophysics. 4th Edition, Chapman and Hall, London, Vol. 360, 104-116**.
- **Reynolds, J. M. (2011). An Introduction to Applied and Environmental Geophysics. Preview, 2011, 33-40. https://doi.org/10.1071/PVv2011n155other**.
- **Sheriff, R. E., & Geldart, L. P. (1995).** Exploration seismology (Vol. 9). Cambridge university press.
- **Telford, W. M., Geldart, L. P., & Sheriff, R. E. (1990).** Applied geophysics (2nd ed.). Cambridge University Press.
- **Rubin, Y., & Hubbard, S. S. (Eds.). (2005).** *Hydrogeophysics*. Springer
- **Rabczak, V., McClymont, A., Hoppe, W., Alexa, Z., & Shih, E. (2023).** Near surface geophysical investigation for geotechnical geohazard assessment using seismic refraction and resistivity survey. *GeoConvention 2023*, Calgary, AB, Canada.
- **Jin, D. J. (1992)**. A review of magnetic induced polarization with a special reference to its applicability to environmental restoration problems.
  Journal of Hazardous Materials, *30*(1), 39–58. https://doi.org/10.1016/0304-3894(92)87073-O

- **Yilmaz, Ö. (2008).** Seismic data analysis: Processing, inversion, and interpretation of seismic data (2nd ed.). Society of Exploration Geophysicists.
- **Vaudelet, Pierre & Schmutz, Myriam & Pessel, M. & Franceschi, Michel & Guérin, Roger & Atteia, Olivier & Blondel, A. & Ngomseu, C. & Galaup, Serge & Rejiba, Fayçal & Bégassat, P.. (2011).** Mapping of contaminant plumes with geoelectrical methods. A case study in urban context. Journal of Applied Geophysics - J APPL GEOPHYS. 75. 738-751. 10.1016/j.jappgeo.2011.09.023.
- **Xiao-qiang, C.. (2006).** Application of integrated geophysical exploration to find concealed ores in the haxi gold mining area. Geology and Prospecting.117–136.
- **Zoeppritz, K. (1919).** On the reflection and transmission of seismic waves. Gott Nachr 1 (5), 66-84.

Index

www.ingramcontent.com/pod-product-compliance
Lightning Source LLC
LaVergne TN
LVHW082344200926
848481LV00012B/3443